# INTRODUCTION

Science is an endeavour to interpret and understand every aspect of Nature, from galaxies to germs and beyond. With the invention of microscopes and telescopes in the 16th century presenting windows to alien worlds at the smallest and largest scales, our horizons expanded rapidly: we discovered that disease was caused by microscopic organisms too small to be seen by the human eye alone, we collected light that had travelled millions of years to reach our planet, giving insight into the evolution of stars, planets and galaxies. Since then our horizons have expanded at a rate not much slower than the speed of light. Scientists have continually found ways to go that extra bit further in deducing what lies beyond our current understanding.

## QUANTUM SHOCK AND AWE

Quantum physics was born out of curiosity for the unexplained and over the past 150 years we have delved inside the atom. What we have found is a world so strange that no human conceived of its existence. Like Alice falling down the rabbit hole to discover Wonderland we have

# CONTENTS

$$\sigma_x \sigma_p \geq \frac{\hbar}{2}$$

$$\boxed{E = h\nu} \quad E = \frac{\hbar^2 k^2}{2m}$$

$$+ A_- e^{-ik_1 x}) \quad x < 0$$

$$+ B_- e^{-ik_2 x}) \quad x > 0$$

$$\frac{d}{dt} A(t) = \frac{i}{\hbar}\left[H, A(t)\right] + \frac{\partial A(t)}{\partial \tau}$$

$$T|j, m\rangle \equiv |T(j, m)\rangle = (-1)^{j-m}|j, -m$$

$$AB = \sum_{i,j} c_{ij}|i\rangle A \otimes |j\rangle B$$

$$H_n(x) = (-1)^n e^{x^2} \frac{d}{dx^n}(e^{-x^2})$$

# BRAINS EXPLAINS
## QUANTUM PHYSICS

CASSELL ILLUSTRATED

An Hachette UK Company
www.hachette.co.uk

First published in Great Britain in 2015 by
Cassell, a division of Octopus Publishing Group Ltd
Carmelite House
50 Victoria Embankment
London EC4Y 0DZ
www.octopusbooks.co.uk

ISBN 9781844038299

A CIP catalogue record for this book is available
from the British Library

Written by Dr Ben Still, physicist and science
communicator, currently Honorary Research Fellow
at Queen Mary University of London
www.benstill.com

Edited by Julian Flanders
Pages designed by Craig Stevens
Artwork by Patrick Mulrey and Craig Stevens
Printed and bound in China

10 9 8 7 6 5 4 3 2 1

been exposed to a fantastical place where our day-to-day understanding does not apply. Because of this, or the image of horrific mathematical equations, the phrase 'quantum physics' may daunt you. Indeed, one of the founding fathers of the science, Niels Bohr, said, 'Anyone who is not shocked by quantum theory has not understood it'. It certainly requires one to leave preconceptions at the door and to accept experimental evidence as proof – two ideal character traits of a true scientist. If you do approach the subject with trepidation, I hope these few pages will change that into shock and awe. I also hope that you will discover something new or see old knowledge in a new light.

These few pages are but a scratch on the surface of the many amazing phenomena that come under the umbrella of quantum physics. Affecting most areas of modern life and providing the greatest international scientific collaborations of all time, our world would be very different if not for its discovery. The greatest result I could hope for is that after reading this book your journey continues further down the rabbit hole to explore in more detail what we cover. But for now let's go into the quantum realm, through history, analogies and artworks to discover new things.

'AFTER BECOMING AN ORPHAN MY ADOPTED FATHER, A CAMBRIDGE UNIVERSITY PROFESSOR, SCHOOLED ME IN THE SCIENCES. **I BECAME FASCINATED FROM AN EARLY AGE AND REVELLED IN LEARNING OF THE BIG THEORIES AND EXPERIMENTS.** ALTHOUGH GENERAL RELATIVITY DESCRIBES THE CONDUCT OF GALAXIES AND BLACK HOLES, **IT IS QUANTUM PHYSICS THAT EXCITES ME MOST.** IN THIS WIDE FIELD THERE ARE AMAZING OPPORTUNITIES FOR NEW TECHNOLOGIES AND UNDERSTANDING OF HOW NATURE WORKS. LET ME TELL YOU A LITTLE NOW OF THIS AMAZING SCIENCE…'

# WHAT IS QUANTUM?

At the turn of the 20th century many physicists thought that our understanding of the Universe was near completion. But, they were ignoring signs that pointed towards some new behaviour of Nature, something a new generation would discover is even more wonderful than science fiction could have predicted.

'FOR DECADES THE WORD "QUANTUM" HAS BEEN USED TO EMBODY CONCEPTS AT THE FOREFRONT OF SCIENTIFIC RESEARCH, JUST AS IT HAS BEEN USED TO GIVE ARTISTIC LICENSE TO ALL SORTS OF STRANGE AND WONDERFUL SCIENCE FICTION. BOTH ARE USES WITH GOOD REASON.'

Since it first emerged in the late 19th century quantum physics had confused, perplexed and humbled many a scientist's world-view. For this reason many established scientists at the time could not accept what experiment was showing them. The reason for this reticence is understandable. Until the end of the 19th and the turn of the 20th century physics had been understood at human scales. Although Sir Isaac Newton's laws, first published

in 1687, described the motion of things under gravity on Earth and could also be used to describe the motion of planets in our solar system, these were things that we could directly see from observing either emitted or reflected light (see page 60).

When we want to understand more about anything we have to push the current horizons of understanding. In quantum terms this was led directly by a pushing of horizons in scale. As Einstein developed his theory of general relativity during the early years of the 20th century

he discovered that at larger than human scales gravity behaved differently from Newton's understanding. When turning our eyes to the smallest scales the differences were even more dramatic. Our everyday experience of the world through science was challenged as experiments led the way into a new and strange horizon. The smooth and predictable nature of Nature was proving to be something else altogether.

## THE QUANTUM SCALE

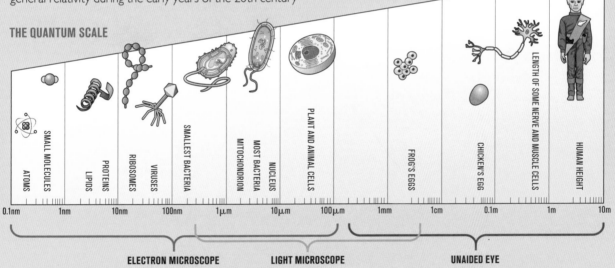

ATOMS
SMALL MOLECULES
LIPIDS
PROTEINS
RIBOSOMES
VIRUSES
SMALLEST BACTERIA
MITOCHONDRION
MOST BACTERIA
NUCLEUS
PLANT AND ANIMAL CELLS
FROG'S EGGS
CHICKEN'S EGG
LENGTH OF SOME NERVE AND MUSCLE CELLS
HUMAN HEIGHT

0.1nm   1nm   10nm   100nm   1μm   10μm   100μm   1mm   1cm   0.1m   1m   10m

ELECTRON MICROSCOPE        LIGHT MICROSCOPE        UNAIDED EYE

# AT ITS SMALLEST, NATURE IS LUMPY

From birth we construct our understanding of the world from our experiences. As humans we think on human scales – for instance, 'I am 183cm tall'. We use microscopes to view things much smaller than us in order to understand the structure of living things. We use telescopes to look into the cosmos in order to understand the motion of the planets. It seems that these scales can continue unabated to a minimum of zero at the smallest scales and the size of the Universe at the largest of scales. It also seems to us that any change in scale is smooth and continuous – like the glassy surface of rippling water.

However, the closer we look at water the less smooth and continuous it starts to look. What looks like pure water might be harbouring bacteria and other microbial life. Even with pure water if we look close enough we see that water is a combination of very distinct units called molecules. These molecules are themselves made up of smaller things called atoms. It is looking like we can keep going, discovering smaller and smaller things. Delving deeper we see that the two

## WATER MOLECULES

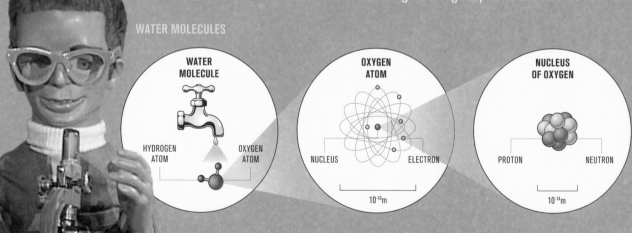

WATER MOLECULE

HYDROGEN ATOM　　OXYGEN ATOM

OXYGEN ATOM

NUCLEUS　　ELECTRON

$10^{-10}$m

NUCLEUS OF OXYGEN

PROTON　　NEUTRON

$10^{-14}$m

hydrogen and single oxygen atoms in each molecule of water are themselves made up of smaller things – so far so good. However, we soon reach a point where we cannot go any smaller. The fact is that there are things in Nature that are not made of smaller things (see page 50).

It seems that it is not just the stuff surrounding us that is lumpy. Energy in its every form also seems to be lumpy when you look at low energy scales. These small lumps are the smallest things it is possible to see

in Nature. So, rather than being smooth and continuous, when looking at the smallest scales, things are broken up into lumps – the same way a sandy beach might look like a flat surface from afar, but is actually made from individual grains of sand. These lumps have been given the name quanta as they represent quantised discrete objects. The science of how these quanta make up our seemingly continuous surroundings and interact with one another is known as quantum physics.

# STARS AND THE BIG BANG
## BLACK BODY RADIATION

'WHEN YOU THINK OF A BLACK CHALKBOARD YOU DON'T IMMEDIATELY LINK IT WITH THE SUN, BUT **THEY HAVE MORE IN COMMON THAN YOU'D EXPECT.**'

A hint that not everything was known about Nature came in 1858 and involved experiments with heated objects. Scottish physicist Balfour Stewart was interested in the radiant heat of metal plates that were either shiny silver or coated in a black layer similar to a chalkboard.

In his experiments he noted that the shiny silver plates were not good at absorbing or emitting, only reflecting thermal radiation, while black surfaces were very good at both absorbing and emitting thermal radiation.

### SPECTRAL RADIANCY

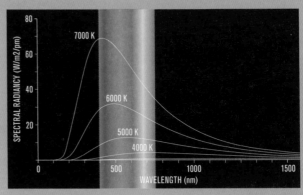

Black body objects at different temperatures emit the most radiation at different wavelengths but the underlying spectrum shape is the same. **IR**

He also discovered that thermal radiation could be reflected and had other properties similar to light. Today we know that thermal radiation is infrared radiation – light just beyond the red in the visible spectrum. If we imagine thermal radiation as light falling on a black surface it makes sense to us that it is being absorbed, as there is an absence of light being reflected back at us.

Absorbing thermal radiation and emitting it back into its surroundings means that any black object quickly blends in with the ambient radiation, taking on a constant temperature dependent on the absorption/emission balance, and reaching what is called thermal equilibrium.

In Heidelberg, German physicist Gustav Kirchhoff independently conducted his own similar experiments, but drew wider reaching conclusions, which he published in 1862. Taking a collection of objects and holding them in thermal equilibrium, Kirchhoff posited that if you measured the thermal radiation being absorbed by and emitted from each object, the ratio of the two at a certain wavelength would be a constant across all the objects. He went further to say that in theory at a given temperature and wavelength this ratio represented an idealised object, which he called a 'black body'.

Of course, his conclusions were empirical and based on trends in experimental data alone. However,

if one knew how this ratio changed with varying temperature or wavelengths then it would provide a deeper understanding of the results. You could also infer the temperature of a black body in thermal equilibrium simply by measuring the amount of thermal radiation it emitted. Although not perfect, stars and planets are in thermal equilibrium and can be modelled as black bodies. The amount of radiation a star emits at different wavelengths can give us a clear picture of its temperature. The temperature of a star is also closely linked to its size, so with just a measurement of the light coming from it we can calculate how big a star is.

## COSMIC MICROWAVE BACKGROUND

The ancient background microwave light is a close real world approximation to an ideal black body. **13**

There are microwaves in every corner of the Universe. These microwaves are ancient light, the afterglow of an event that happened over 13 billion years ago. They have been measured by a number of experiments to show the closest approximation to an ideal black body ever seen. The black body-like spectrum of microwaves showed that at some point in the distant past the entire Universe must have been in contact for there to have been thermal equilibrium – much the same as the gas in a star like the Sun. For this to have happened it must therefore all have come from a common origin – this is the idea behind the Big Bang.

# EXPERIMENT LEADS THE WAY

As empirical experimental evidence continued to gather, a number of unsatisfactory theories appeared. In 1900 Max Planck published a paper, which included a formula that fitted the black body experimental data, but provided no resolution. In the same year Lord Rayleigh and his colleague Sir James Jeans were working on a classical explanation for the data. Their results, published in 1905, showed that it was impossible to describe the emission of light from a black body using the classical wave-like understanding. Although at long wavelengths results showed a minor disagreement between their theory and what was observed by experiment, at small wavelengths the theoretically predicted emission of energy went through the roof – approaching infinity as the wavelength of light approached zero – this disagreement was called the ultraviolet catastrophe.

Another great scientist, Stefan Boltzmann, had shown the relationship between temperature and the energetic vibrations of gas molecules. Planck imagined

## THE ULTRAVIOLET CATASTROPHE

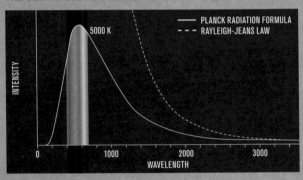

The theory of Rayleigh and Jeans (represented by the dashed line) was not able to reproduce the measured black body spectrum (solid line), blowing up to infinite intensity and energy at small wavelengths. 🆃🆁

that thermal radiation was something similar. He thought of thermal radiation not as a wave but as a collection of vibrating objects. These oscillating units would only vibrate at certain frequencies similar to a guitar string. The string of a guitar is fixed at both ends. If we looked at the lowest frequency note a string can produce in slow motion, we would see a single point of maximum vibration in the middle of the string. Such a vibration would be just half of a full wave. Higher frequency notes are constructed from more half wavelengths and vibration points, in the same way Planck imagined that there was some half wavelength, which defined a minimum unit from which thermal radiation could be constructed. This idea removed the possibility of energy emitted inflating to infinity because thermal radiation could not have a wavelength of zero. This minimum unit of energy depended upon the wavelength of the radiation multiplied by a constant – today this is called Planck's Constant – a fundamental law that has implications for all fields of physics (see page 65).

## GUITAR STRING HALF LENGTH QUANISATION

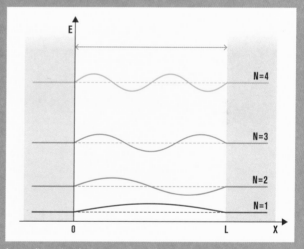

In one dimension the lowest energy of an electron can be represented as the lowest energy vibration of a string, which is just half a full wave (blue). The subsequent more energetic strings are multiples of this lowest half wavelength energy: 2 (red), 3 (green), 4 (orange), etc.

# EINSTEIN'S NOBEL PRIZE

Five years into the 20th century a patent clerk in Switzerland published four papers that revolutionised physics. Having shunned the traditional academic route, Albert Einstein preferred to ruminate on the puzzles he found interesting. He would complete his daily clerical tasks in a few hours, which left him lots of time to sit and think. He conducted thought experiments in which he followed logic and reason to derive outcomes. The most famous of these introduced the world to the special theory of relativity, but it is another of his papers that we will look at here.

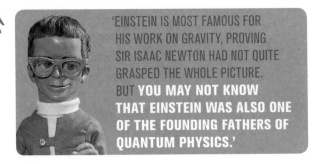

'EINSTEIN IS MOST FAMOUS FOR HIS WORK ON GRAVITY, PROVING SIR ISAAC NEWTON HAD NOT QUITE GRASPED THE WHOLE PICTURE. BUT **YOU MAY NOT KNOW THAT EINSTEIN WAS ALSO ONE OF THE FOUNDING FATHERS OF QUANTUM PHYSICS.**'

In 1887 Heinrich Hertz discovered something quite by accident. He was interested in testing James Clerk Maxwell's theory of electromagnetic waves. Coils of wire were separated by a small gap and one had a large electric current passed through it. When placed close enough together a spark of electricity would cross the gap from the charged coil to the other, much like a tiny lightning bolt. When this spark occurred Hertz detected that radio waves were produced. All of this experimental equipment was built and tested in the sunlight of his lab. To see the spark more clearly Hertz put the apparatus in a dark box, and he noticed something peculiar. In the dark his coils did not spark until he reduced the gap between the coils. In sunlight it seemed easier for a spark to cross a larger gap than in the dark.

## THE PHOTOELECTRIC EFFECT

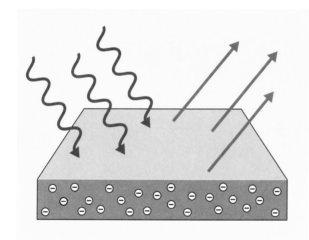

The photoelectric effect in which each photon of light (in blue) strikes an electron with enough energy to release it from the metal (in red). **IR**

Hertz did not pursue this curious behaviour anymore, but others took up the challenge. They charged metal plates and spheres with negative electricity. They noticed that when ultraviolet light was shone on these negatively charged objects they would rapidly lose their charge. They also showed that if one takes an uncharged piece of metal and shines ultraviolet light on it then it becomes positively charged.

In 1896, J. J. Thomson was experimenting with cathode rays – radiation emitted from highly electrically charged metal wires. Using magnetic fields to deflect the cathode rays, Thomson was able to show that they were not made of waves, as other radiation was thought to be, but of tiny particles too small in mass to be whole atoms. The following year Thomson turned his attention to the phenomenon seen by Hertz. He placed a metal plate in an evacuated glass tube and shone ultraviolet light on it. He was able to detect an electric current at some distance from the plate when the light was switched on but not when it was switched off. He concluded that the current was a result of the same cathode ray electron particles being emitted from the surface of the metal. He theorised that the ultraviolet light was vibrating the atoms in the metal until the amplitude became so large that they spat out electrons.

# LUMPY LIGHT:
# THE PHOTOELECTRIC EFFECT

Following J. J. Thomson's experiments, Philipp Lenard discovered that ultraviolet light could be used to separate the neutral molecules of a gas into positive electrically charged ions and electrons. In 1902 Lenard discovered something that was counter intuitive to Maxwell's seemly ever-successful wave theory of light, showing that the number of electrons emitted did not increase with the frequency of the light.

Imagine a sand bank on Tracy Island beach that is gradually being eroded by the tide. If you watch you will note that larger waves with more energy draw more sand away from the bank and out to sea.

In Maxwell's wave theory of light the same would be true for electrons in a metal. The higher the energy of the electromagnetic wave the more electrons (sand) you would expect to see liberated from the metal (sand bank). But what was seen in experiments did not agree with this. If the energy of light is increased by increasing its frequency, it does not increase the amount of electrons released. Instead it simply increases the energy of the electrons that are released.

To increase the number of electrons released it seemed that you had to increase the intensity of the light rather than the frequency. More light meant more electrons released and higher energy of light resulted only in a higher energy of electron being released. This is not at all like the sand being washed away by a wave. Albert Einstein (pictured here) solved the puzzle in a paper called 'The Photoelectric Effect'. Instead of thinking of light as a wave, he thought of it as small packets of energy. These small packets, or quanta, are called photons. The idea for quanta was first put forward by Max Planck in his understanding of thermal radiation (see page 12). These particles of light have a certain energy, which can be increased by increasing the frequency of light. You can increase their energy but at the end of the day there are still only the same amounts that can possibly bump into an electron and set it free. If we increase the intensity of light, however, we increase the number of photons. More photons mean that they can bump into and free more electrons.

Einstein's theory cemented the idea of light and particles as discrete quanta lumps. He had provided the solid ground on which all of quantum mechanics could then be built and earned himself the Nobel Prize in Physics in 1921. Much to his dismay his findings flew in the face of his world-view, bringing with them a revolution in the way we have to think of Nature at its smallest scales.

# CHEMICAL FINGERPRINTS

The word atom is derived from the Greek *atomos*, which means 'indivisible'. The view that everything was made from these tiny units of matter had held since ancient Greek philosophers first expressed it, but the discovery of the electron showed the existence of smaller things within the atom.

Experiments conducted by J. J. Thomson found that electrons had much smaller mass than any atom, and that unlike atoms (which have no charge) electrons possessed a negative electric charge. To balance the negative charge of the electrons an atom must have some subatomic structure that had positive charge.

## PLUM PUDDING

Thomson's theory imagined the atom to be like a plum pudding – a sea of positively charged cake, interspersed with negatively charged electron fruit lumps. Inside each atom there would need to be a fine balance in the distribution of negatively charged electrons. Though his ideas would benefit the modern field of quantum computing (see page 89), Thomson's model was proved to be inconsistent with Nature just five years later.

In Manchester, Hans Geiger and student Ernest Marsden conducted groundbreaking experiments under the watchful eye of Ernest Rutherford, which aimed to test if the atom was, indeed, like a plum pudding. Some years earlier Rutherford discovered positively charged particles emitted in the radioactive decay of heavy atoms. This alpha radiation (ions of helium) was the probe to look inside larger atoms. Alpha radiation was fired at thin foils of gold and the paths they took were measured. If gold atoms had an even distribution of electric charge, as in Thomson's pudding, then you would expect most of the particles to pass straight through, their paths possibly being deviated slightly by a tiny imbalance in the distribution of negatively charged electrons. However, something altogether different was observed.

Although most of the alpha particles passed right through, there were some that scattered back in the direction they had come. The only way that this could occur, Rutherford suggested, was if the positive charge within an atom was not evenly distributed. If all the positive charge was located in a small space it would be able to deflect the path of the alpha particles, as seen in the experiment. This shift in understanding meant that, in fact, an atom is mostly empty space, with a nucleus of positive charge surrounded by a thin spread of electrons.

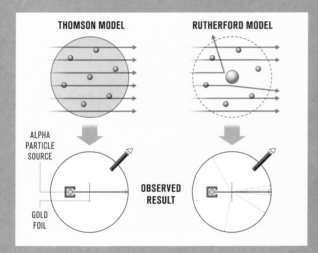

THOMSON MODEL     RUTHERFORD MODEL

ALPHA PARTICLE SOURCE

OBSERVED RESULT

GOLD FOIL

If the atom were like a plum pudding (left) then the alpha particles in Rutherford's experiment would not deviate from their path. Alpha particles were observed in some cases, however, to be deflected at large angles, which may only be explained if there was a high concentration of positive charge within the atoms (right). **IR**

The planetary atom: a central positively charged nucleus containing most of the mass of the atom, orbited by a number of low mass negatively charged electrons. **IR**

# THE PLANETARY MODEL

In May 1911 Rutherford summed up the state of atomic knowledge at the time. In his paper he suggested that the central charged nucleus of any given atom of an element, discovered by Geiger and Marsden, might be linked in some way to the position of that element on the periodic table. He stated, 'the central charge in an atom is approximately proportional to its atomic [mass]'. The atomic mass is the mass of a single atom of an element, counted as a multiple of the mass of a hydrogen atom: carbon has an atomic mass of 12, a mass 12 times greater than that of hydrogen. The periodic table arranges elements in order of increasing atomic mass. Rutherford did not make a solid prediction of this relationship instead remarking simply, '[i]t will be of great interest to examine experimentally whether such a simple relation holds'. It was Antonius van den Broek who just a month later explicitly predicted such a connection between atomic mass and charge of the nucleus. Within two years it was Englishman Henry Moseley who proved the link and provided chemistry with a firm fundamental basis in atomic physics.

When it came to the electrons within the nucleus Rutherford built upon theories of other eminent scientists, putting forward in his paper the idea that the electrons orbited the nucleus like small satellites around a planet. This was an idea first suggested by Japanese physicist Hantaro Nagaoka and based upon the stability of Saturn's rings. Small debris of rock and ice are bound to orbit as rings around Saturn in the same way as the comparatively lightweight *Thunderbird 5* is bound into orbit by the gravitational pull of the much more massive Earth. Nagoka imagined that the low mass negatively charged electron satellites orbit a much more massive positively charged central nucleus. After the results of the Geiger-Marsden experiment this seemed to be a sound logical step.

## ELECTROMAGNETIC ATTRACTION

The planetary model, as it became known, assumes two things: firstly that the electrons are far less massive than the central nucleus of the atom, and secondly that the attraction between the electrons and the positively charged centre is the same magnitude as that attracting the rings to Saturn. The first assumption holds, thanks to Thomson's measurement of the small electron mass with respect to that of an atom. It is in the second assumption, however, that the planetary model falls down. Although it may not seem so day-to-day, gravity is a very feeble force – by simply jumping up we can overcome the gravitational force exerted by an entire planet. The force of attraction between electrons and the nucleus on the other hand is much stronger. The opposite electric charges are drawn to one another through the electromagnetic force, which is many times stronger than the force of gravity. The superior strength of the electromagnetic force can be demonstrated through the ease with which one can pick up metal with a bar magnet, overcoming the Earth's gravitational pull in the process. Greater attraction between these electrically charged particles means that orbits of satellite electrons should not be stable for long inside an atom. Electrons, attracted by the positive electric charge, should spiral into the nucleus in a fraction of a second. The fact that there are stable atoms everywhere meant that there must be more to the story; again it is quantum physics that provides an explanation.

# FINGERPRINTS

In the 1880s Johannes Rydberg was investigating the light emitted by different gases. Scientists before him had noticed that gases emitted light at specific wavelengths. These emissions built up a unique pattern of spectral lines from which each gas could be identified. Some fifteen years before, French astronomer Jules Janssen saw a pattern of bright spectral lines while looking at a solar eclipse. Later that year Joseph Lockyer saw the same emission and agreed with Janssen that it was an unknown element. The first chemical element ever discovered on a planet other than the Earth, it was named helium, after the Greek Sun god Helios.

Rydberg set about looking for patterns in these fingerprints. He realised he could simplify his calculations greatly with a mathematical trick. Instead of looking at the wavelength of the light, he looked instead at something called the wavenumber. Just inverse of the wavelength, the wavenumber represents the number of waves that exist in some unit of length. For instance, if a radio wave had a wavelength of 0.5 metres then its spectral wavenumber would be 1/0.5 = 2 per metre. He numbered the spectral lines 1, 2, 3… in order of increasing wavenumber (decreasing wavelength) and for

**HYDROGEN**

**HELIUM**

Elemental fingerprints, showing the energies of light each can emit. **IR**

each plotted them against their calculated value. When he looked at the results for a number of different gasses it was obvious that they each showed the same pattern.

Rydberg then looked for an empirical mathematical formula that fitted the pattern. He noticed a formula written by Johann Balmer, which accurately described the spectral lines of hydrogen gas. Rydberg re-wrote the equation in terms of wavenumber and stumbled across a fundamental connection between the wavelength of light and its energy. It showed that light emitted from an atom was linked to an arrangement of energy levels within it. This same law could be applied to a number of atoms, suggesting they had a common structure. As with much of quantum physics, the interpretation of this law forced scientists to re-think their classical view of the world.

# THE EXPANDING UNIVERSE

Every different type of chemical element has a unique spectral line fingerprint. When looking at the spectral lines of light from the Sun and other stars we can see that they are made primarily from hydrogen gas. We can also tell the composition of a planet's atmosphere from looking at the light it absorbs. Most interestingly we can use the fingerprints to measure the expansion of our Universe. Because of the abundance of hydrogen in stars we know how to find its unique fingerprint everywhere in the night sky. When looking at distant galaxies we see that this fingerprint shows the same pattern, but that the wavelength of each spectral line has increased. This occurs as the light from the galaxy is stretched as is moves away from Earth – the faster it recedes the greater the change in wavelength. This is how, in the 1920s, Edwin Hubble determined that the Universe we live in is expanding at an ever-increasing rate. The further a galaxy is from Earth the further towards longer wavelengths the spectral lines were observed to be shifted. Because this makes the light redder in colour this is known as redshift. Without this understanding of the tiny world of quantum physics we would not have been able to investigate the largest scales in Nature.

## ABSORPTION LINES

**FROM OUR SUN**

**FROM A SUPER CLUSTER OF GALAXIES**

The fingerprint of chemicals found in a star from the energy of light they absorb. When not moving these would be near identical, but light from the distant cluster of galaxies has been shifted towards red light. ⏴⏵

# ELECTRONS IN A BOX

Niels Bohr's interpretation of experimental results from his predecessors cemented the idea of quanta within the atom. He took Rutherford's idea of a planetary-like atom and married it with Rydberg's empirical understanding to produce the first quantum model of the atom. His genius was to connect the light emitted from atoms to the orbitals of the electrons. His idea was that electrons had well defined orbits determined by their wavenumber and that the light emitted by an atom was the energy released when an electron passed from one orbit to another.

If we first imagine electrons not as J. J. Thomson's small lumps, but instead like a wave. An electron can then only exist as a vibrating wave. We should imagine the orbitals of an electron as a number of vibrating strings in a box. The end of a string is securely attached to the box and therefore cannot vibrate. If we pluck the string we set it in motion, vibrating. The most simple of vibrations is where there is a single point in the middle of the string, which is vibrating up and down. Let us compare this to a full wave on water, which has both an upward peak and a downward trough. Our simple vibrating string can only be thought of as half a wave as it has either one peak or one trough at any one point in time. It is possible to create more complicated waves, as long as each end of the string is fixed. The next more complicated wave has two peaks or troughs. This would be a full wave. We can continue in this fashion, adding half a wave each time to create ever more complex waves. The vibration of the strings in the box are quantised and built from some multiple of the smallest lump (quanta), which is a half wavelength.

If these waves were on a piece of string you would need to move the string faster to produce more

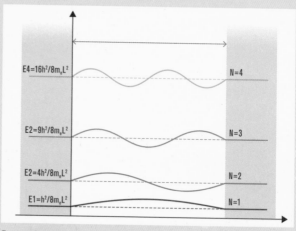

$E4 = 16h^2/8m_eL^2$     N=4

$E2 = 9h^2/8m_eL^2$     N=3

$E2 = 4h^2/8m_eL^2$     N=2

$E1 = h^2/8m_eL^2$     N=1

Energy of electrons in a box. The lowest energy is the least energetic vibration and each subsequent level an addition of this smallest quanta of energy. ⒷⓇ

complicated waves. This would require more energy from the person moving/plucking the string. Indeed this is exactly what these different waveforms represent: the more complicated they are the higher energy the possible electron orbital they represent within an atom. When taking energy from the string by absorbing its movement in some way we might go from a complicated wave to a simpler one. This is exactly how light is emitted from an atom, the same definite wavelengths of light in which Rydberg noticed patterns. When an electron orbiting an atom falls down to a lower energy orbital, the difference in energy is released in the form of light.

Interpreting electron orbitals like this provided answers to some of the biggest questions left open by the planetary atomic model. The most pressing was why the electrons do not forever spiral in towards the nucleus of the atom to which they are attracted. Classically, in the planetary model of the atom, we can think of the nucleus as a valley and the electron as a ball. The same way a ball would roll down to the bottom of a valley under gravity, a negatively charged electron would be drawn towards the central positively charged nucleus. Here, both electron and ball have zero potential energy. In quantising the energies of electrons in an atom they can never reach zero energy and can therefore never reach the nucleus. This is because the energy any orbiting electron can have must be some positive multiple of the minimum quanta. The lowest energy an electron can have is a called the ground state; it has one peak or one trough, it is half a wave.

Bohr had showed the way to a stable atom through quantisation. As experiments looked with more scrutiny at atoms they continued to uncover new questions that led physics further down the quantum road.

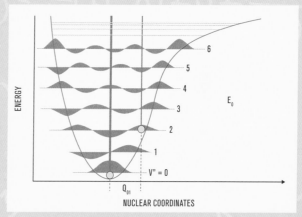

The energy of electrons within an atom. Classically the electron can be thought of as a ball, which would roll to come to rest at the very bottom of a valley, careering into the nucleus. The lowest energy possible for the quantum electron is not zero and so keeps it from ever being able to fall into the nucleus. 👓

# MODERN CHEMISTRY
## QUANTUM PHYSICS OF ELECTRONS

The 1920s saw what became known as old quantum theory left behind and a new quantum theory emerge. One of the catalysts was graduate student Louis de Broglie whose thesis, published in 1924, combined the idea of electrons as particles *and* waves. His solution showed that each particle had an associated wavelength dependent on the inverse value of a certain property of the particle – its momentum.

The momentum of any object is the mass of the object multiplied by its speed in a particular direction. Once you know the momentum of an object then you can calculate its de Broglie wavelength. As the wavelength depends on the inverse of the momentum, a large momentum would give a small wavelength and visa versa.

Every object with a momentum has a de Broglie wavelength, even you and me. Human beings appear a solid lump, but are made from a collection of trillions of subatomic quantum particles. Because of this we generate trillions of times more momentum than a single quantum particle travelling at the same speed. This large momentum gives us a tiny de Broglie wavelength, so small that it is practically impossible to observe any of the wave-like properties of a human. Tiny quantum objects have much smaller masses and therefore smaller momentum. This leads to larger de Broglie wavelengths, which are much more likely to be observed in experiments. The wave-like nature of electrons was subsequently proven in an experiment conducted by Clinton Davisson and Lester Germer, in which they showed electron waves interfering with one another as they spread out like ripples from multiple stones dropped in a pond (see page 47).

De Broglie's idea was the beginning of modern quantum physics (though the phrase itself was not used until 1931). Although working independently, Erwin Schrödinger and another group of scientists, Werner Heisenberg, Max Born and Pascual Jordan, evolved de Broglie's ideas into a general theory. However, Schrödinger's wave mechanics and the matrix mechanics of Heisenberg, Born and Jordan, only worked when the speeds of the particles in question were low, less than the speed of light. It was actually Paul Dirac who evolved Schrödinger's ideas into a theory that could be applied no matter the speed, as de Broglie had originally intended.

The fact is that a particle with a small mass only requires a small amount of energy to move fast. If you were to throw a tennis ball and a bowling ball with the same energy, for example, the tennis ball would travel faster. Tiny quantum particles therefore tend to move at very high speeds, large fractions of the speed of light.

At these speeds, Einstein told us in his final paper of 1905, things behave very differently. The speed of light in empty space is the upper speed limit of the Universe; nothing can travel faster. To enforce this speed limit, the theory of relativity states that the mass of an object effectively increases the closer it gets to the speed of light. The faster an object travels the more massive it becomes, becoming infinitely massive at the speed of light. This mass increase, in turn, increases the energy required to change its speed. It would require an infinite amount of energy to bring anything with a mass to the speed of light. Once Dirac included this speed limit in Schrödinger's equations (along with some ideas from Wolfgang Pauli, see page 64) he had constructed a mathematical explanation of de Broglie's particle waves that could be applied to the widest array of situations possible.

**SPEED OF LIGHT** 671 million mph (186,282 mps)
**Time taken to travel from the Sun to Earth** 8.3 minutes
**SPEED OF *THUNDERBIRD 3*** 25,200 mph
**Estimated time taken to travel from Earth to the Sun** 65 hours

Dirac's equation for quantum objects brought with it, for the first time in quantum physics, theoretical predictions. The theory suggested that the electron had a polar opposite, a particle that to date had not been seen in experiment. The negatively electrically charged electron was predicted to have an opposite positive electrically charged version of itself with the same mass. Dirac had predicted the existence of new particles, later called antimatter. A few years after Dirac published his paper, these antimatter positrons were seen to exist in cosmic rays (showers of charged particles that rain down upon us constantly). This was the first time that theory had led experiment into the unknown in quantum physics.

# ELECTRON ORBITALS

Today the field of theoretical chemistry is founded firmly on quantum physics. What we once thought of as a mini solar system of electrons orbiting a central nucleus has turned into something altogether more strange. Modern chemistry does not describe the energy levels of electrons in an atom through their wave-like behaviour alone, nor directly as vibrating strings. When talking about strings earlier we were considering one-dimensional up and down vibrations.

Although Bohr's initial thought experiments led us to a great understanding, atoms are actually three-dimensional. The more dimensions we consider the more complex patterns can be formed. In two dimensions we go from

## HYDROGEN WAVE FUNCTION

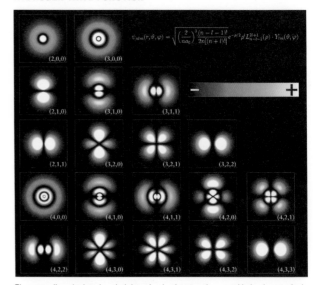

Theoretically calculated probability clouds showing the most likely place to find electrons of different energies surrounding the hydrogen nucleus. The increasing complexity of these clouds follows an increase in energy.

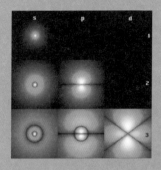

a simple wave pattern to those that look like a snowflake. In three dimensions we need to think of them as clouds. The wave-like nature of electrons suggests that they do not certainly exist in definite locations (see page 38). Instead electrons have a probability to be seen at different locations surrounding the atomic nucleus. If one were to draw a boundary around where the electrons were most likely to be seen (say 90 per cent of the time) then they would form a pattern of clouds. Their size, shape and location around the nucleus define the energy of the electrons that are found there. As with one-dimensional waves, the simpler a cloud looks the lower in energy it is. As one goes higher in energy the clouds become more complicated in pattern.

## THE HYDROGEN ATOM

Let us take hydrogen as an example atom. The lowest energy an electron can have is a simple spherical cloud surrounding the nucleus; this is akin to the simple half wavelength of the one-dimensional wave. If an electron were to gain energy, and therefore momentum, its three-dimensional wave pattern would change and become more complex. Similar to the increased complexity of a one-dimensional wave there would be an increased fragmentation of the cloud. The next most complex cloud is simply split into two. As the electron here is on average more likely to be held further away from the nucleus it must have a higher potential energy, as a ball held further up a hill would have more energy. As we go to the next energy level things become even more complicated. The ever-increasing complexity is each time pushing the average location of electrons away from the nucleus – even higher up the hill – the further it goes the more potential energy it possesses.

# CHEMICAL BONDS

A chemical bond is a mutually beneficial coming together of two or more atoms to form a molecule. Everything in the Universe naturally desires to be at the lowest possible energy possible: high energy means highly unstable; low energy means stability. Chemical bonding is seen in quantum physics as a change in the shape of the orbitals to lower the energy of the electrons within atoms. These bonds which bind atoms together into molecules are formed either through the sharing or exchange of electrons. In exchanging electrons the atoms involved each acquire opposite electrical charges. These opposite electrically charged atoms are attracted to each other in what is known as ionic bonding.

The sharing of electrons results in what is known as covalent bonding. Sharing reduces the energy of the electrons involved, which forms a more stable whole than a single atom. This can be viewed in a number of ways depending upon your favourite analogy used so far. When thinking of the electron waves in a box: the energy is reduced because the sharing of electrons stretches the width of the box, increasing the minimum wavelength, which decreases the energy. When thinking of electron orbital clouds: the sharing increases the likelihood that an electron is in the region between bonding atoms, drawing them closer to the nucleus. Either way a reduction in energy is seen when two or more atoms form chemical bonds.

Schrödinger's wave theory describes the electrons in atoms, but was first used to calculate the energy levels seen in molecules of hydrogen gas. Pairs of hydrogen atoms bond together to form hydrogen gas molecules, $H_2$. Walter Heitler and Fritz London showed in 1928 how the wave-like behaviour of two hydrogen atoms combined to form a molecule of hydrogen gas. Their idea was that the electrons were shared between the two atoms to generate a newly formed cloud localised between the two bonding atoms. This cloud was akin to a chain linking two atoms directly and filling the space between the atoms. This idea was further developed by theoretical chemist Linus Pauling into valence-bond theory, which draws intuitively from classical chemistry.

## MOLECULAR ORBITAL THEORY

A short time later Friedrich Hund and Robert S. Mulliken developed a competing theory. They too understood bonding to be the sharing of electrons, but thought that the sharing was not localised between just the bonding atoms. Instead they predicted that the electrons involved in bonding were delocalised; their wave/cloud spread out across the entire molecule. In this way the electrons formed not a chain between the bonding atoms but more of an elastic wrapper that pulled all atoms in the molecule together. This came to be known as molecular orbital theory, as one could imagine the electrons orbiting the nuclei in the entire molecule rather than a single atom.

As with all science, the proof was in the experimental results. Physicists and the first quantum chemists looked at the spectrum of light coming from molecules. The energies they measured determined the energy levels of the electrons, which were then compared with those predicted by the two theories. Although classically the least intuitive, molecular orbital theory provided the most accurate explanation.

---

Chemical bonds are formed between two atoms through the sharing (top) or exchange (bottom) of electrons. The sharing is known as covalent bonding and the exchange ionic bonding. **IR**

## COVALENT/IONIC BONDS

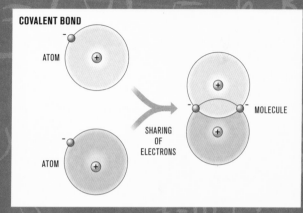

**COVALENT BOND**

ATOM

ATOM

SHARING OF ELECTRONS

MOLECULE

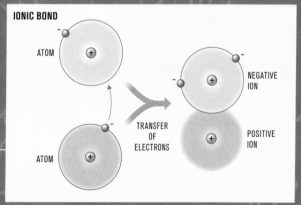

**IONIC BOND**

ATOM

ATOM

TRANSFER OF ELECTRONS

NEGATIVE ION

POSITIVE ION

# THE EXCLUSION PRINCIPLE

From the late 19th century scientists began to believe that atoms with an even number of electrons seemed more chemically stable than those with an odd number. This was evident when atoms combined to form molecules; for example, single electron hydrogen atoms do not exist alone but instead combine as a molecule sharing two electrons.

'WOLFGANG PAULI KNEW OF THE ATOMIC PREFERENCE FOR EVEN NUMBERS OF ELECTRONS AND MADE IT HIS TASK TO UNDERSTAND THE RESULTS OF ZEEMAN'S FINDINGS.'

This puzzled many, as there did not seem any obvious reason for it. In search of an answer scientists looked once more at the light emitted from atoms. Dutch physicist Pieter Zeeman placed his light source inside a magnetic field and then looked at the spectrum of light. Where there had previously been single well-defined spectral lines most seemed to split into two distinct spectral lines surrounding the original position of the unmagnetised single line. When the results were presented at a conference in 1896, Hendrik Lorentz provided an explanation. He explained that the light emitted came from the vibration of negatively charged

particles thousands of times lighter in mass than the atom. This was the first real evidence of electrons and the realm of the subatomic a year before J. J. Thomson's discovery.

## THREE DIMENSIONS +?

Up to this point, early 1920s physicists had been describing the energy levels of electrons using a system of three numbers labelled n, m, and l. These three quantum numbers represented the three dimensions of space that the electrons orbited in. This idea is an extension of the one dimensional string in a box (see page 24) into three dimensions. In one dimension we have just one number that denotes the number of quanta that make up that energy level, n. In looking at the world in three dimensions we need numbers for the other two dimensions: m and l. These three numbers were able to explain and predict the energy levels of electrons seen in experiment and led to the quantum understanding of the atom. But Pauli realised that if you desired additionally to explain the splitting of spectral lines seen by Zeeman then you would have to add another quantum number to the mix. Pauli assigned a new quantum number, which had just two values to provide the newly seen splitting.

### THE ZEEMAN EFFECT

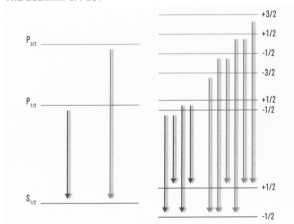

The energy of light emitted by an atom is defined by the difference in energy levels electrons in an atom can occupy. When a magnetic field is applied to an atom these levels are seen to split, a phenomenon known as the Zeeman Effect. **IR**

# YOU SPIN ME ROUND

Although Pauli's extra quantum number provided a mathematical solution for the splitting of spectral lines observed in experiment, it was not known what this number meant physically. It was suggested that the number represented the fact that an electron is itself rotating as it orbits an atom. In a similar way to the revolving Earth,

a revolving electron would produce a magnetic field through the electromagnetic force (see pages 52–3). A spinning electric charge produces a magnetic field similar to a bar magnet. Just as a bar magnet has a north pole and a south pole, there would be a defined direction of the magnetic field of a spinning electron. The direction and size of the magnetic field is known as the magnetic moment.

The magnetic moment of an electron in an atom comes from two sources. The first is the rotation about itself as we have discussed; the second is the motion of the electron around the positively electrically charged atomic nucleus, captured in the quantum number m. The portion that arises from its own spinning shows rather strange behaviour. Quantum objects like electrons are not understood as solid physical objects. They are known to behave like waves and particles (see page 49). To say, therefore, that an electron spins does not make physical sense; we cannot imagine it as a spinning top on a table. The idea of spin in the quantum world is an internal property that arises from the mathematical explanation of how particles are seen to behave when observed. If we did try to model an electron classically as a small ball with an electron's electric charge we would find that it did not generate the same size magnetic field as an electron. In fact electrons, and particles like it, create magnetic fields

## ELECTRON SPIN

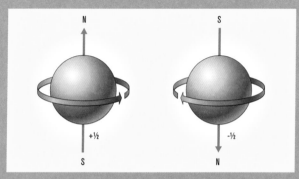

An electron's quantum spin can be said to be up (left) or down (right).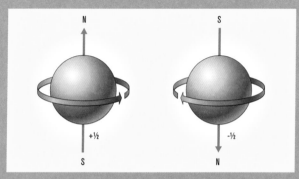

two times stronger than expected classically. This clearly shows us that quantum spin is altogether different.

## QUANTUM SPIN

The electron and all other particles that build up atoms around us are said to have a spin of ½. If we were to think of quantum spin in geometry, spin ½ particles would appear very strange indeed. Turning any everyday item around on itself once by 360° will leave it in exactly the same position – this is called rotational symmetry with order 1, as the item looks the same once every 360°. A square could be said to

have rotational symmetry of order 4 because there are four times in a 360° rotation that the square would look indistinguishable from its start point. Rotational symmetry of order ½ means an electron would have to be turned around not once but twice, a full 720°, before it looked the same as the position it started in.

Quantum numbers cannot be measured directly but are instead determined from experimentally measurable quantities. As this is quantum physics the magnetic moment cannot take just any value and must be made up of an addition of quanta. The smallest quanta block of magnetic moment is known as the Bohr magneton. Pauli's quantum spin number is multiplied by the Bohr magneton and another number known as the g-factor to give the experimentally determined magnetic moment. The g-factor, which equals 2, accounts for the amazing ability for quantum electrons to be able to produce a stronger magnetic field than could be achieved classically. Our understanding of particles and the forces that bind them make very accurate predictions (1 part in 400 billion) about the value of the g-factor. Experiments around the world are using these predictions to expand our understanding of the particles. Any small deviation from the value predicted would suggest that there is some flaw in our understanding of Nature.

# EXCLUSION'S INCLUSION

All of this led Pauli to his greatest work – the exclusion principle. In it Pauli states that no two electrons (or similar particles) can occupy the same quantum state/ numbers. This meant that although there are a number of different energy levels within an atom, not all the electrons are sitting in the lowest energy state, closest to the nucleus (the way a ball would come to rest at the bottom of a valley). The different energy levels defined by the three numbers n, m, and I could each have two electrons in them – each with different spin. But as soon as two electrons occupied that energy level there would be no choice for other electrons but to occupy a higher energy level. We now understand that the chemistry of the elements is determined by the number of electrons that an atom has in its outermost shell. Using these four quantum numbers chemists could give physical meaning to the patterns seen in the periodic table of elements constructed by Dmitri Mendeleev in 1869.

The exclusion principle is also responsible for atoms having any such physical size. After Rutherford discovered that the majority of an atom was in the nucleus, some 100,000 times smaller than an atom

seemed to be, then there needed to be something to fill the rest of the atom. Filling the energy levels and obeying the exclusion principle provided an explanation as to why not all the electrons were as close to the nucleus as they could be, and therefore why atoms appeared so large.

The exclusion principle also partly explains why you do not fall through your chair. While it is mainly the repulsion between the negatively charged electrons in your behind and those in the chair being sat on, it is also due to the fact that those electrons cannot and will never share the same space.

## BLACK HOLES

Perhaps the most amazing example of the exclusion principle at work is found in astrophysics; there are entire stars who owe their stability to it. When a star dies and

light is no longer produced to push against it,
gravity wins the fight. All the materials that make
up the star are squished into one another under
their own weight. In the largest of stars this pressure
can be so great that it deforms the shape of the
atoms in the material. White dwarf stars are
supported from collapsing any further from the
pressure caused by electrons in atoms having
nowhere else to go. They are compressed into the
lowest energy levels they can be in and the atoms
are said to be degenerate. The exclusion principle
means that they push against gravity and this pushing
is known as degeneracy pressure. If a star is much
larger the force of gravity can press the electrons
to merge with the positively charged protons in the
centre of every atom to form neutrons. Neutrons
have quantum spin of ½ like electrons and so they
too obey the exclusion principle. These larger
particles can produce a larger degeneracy pressure
and stave off more gravity. If a star is any larger then
gravity will win out entirely against quantum physics
and a black hole is formed. This is where quantum
physics meets Einstein's general theory of relativity
and where our understanding of Nature as described
by them breaks down.

# ALL IN THE ROLL OF A DICE
## POSSIBILITY AND PROBABILITY

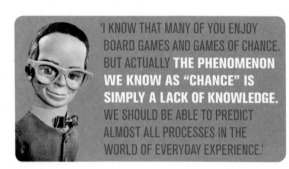

'I KNOW THAT MANY OF YOU ENJOY BOARD GAMES AND GAMES OF CHANCE. BUT ACTUALLY **THE PHENOMENON WE KNOW AS "CHANCE" IS SIMPLY A LACK OF KNOWLEDGE.** WE SHOULD BE ABLE TO PREDICT ALMOST ALL PROCESSES IN THE WORLD OF EVERYDAY EXPERIENCE.'

At quantum scales there is no definitive answer or outcome to anything. Instead there's a whole host of possible answers, each with its own probability. Einstein commented on quantum physics in reference to God: 'I, at any rate, am convinced that He does not throw dice.'

Einstein offers a brilliant analogy of a dice as a quantum object. When a dice is rolled it is not certain which number it will actually land on – board games would be very boring if it were. However, if you were a gambler you could predict the probability of rolling each number. There are six sides to a dice and so one would expect, if the dice is fair, that the probability of rolling 1, 2, 3, 4, 5, or 6 would be $\frac{1}{6}$, or 16.67 per cent. Before rolling, if there is one spot on the top face of the dice we can say that for certain that it has a score of 1. We then pick up the dice, shake it a little and roll it. When the dice is shaking and rolling it no longer has a definite score, instead it is a mixture of all the six possible figures, each having a probability of $\frac{1}{6}$ to be the final answer. Only when it stops rolling can we see which of the numbers is facing up. A dice in motion is just like a quantum object in motion that is not being looked at. Between the start and finish of some experiment a quantum object is just a mixture of possibilities, each with their own probability.

# LOADED DICE?

If we suspected that the dice was loaded and wanted to test it, we could run a simple experiment. We roll the same dice many, many times. Each time it comes to rest we note down the score. When we have done this lots of times we tally up the number of rolls that gave each of the six scores. We then divide the results by the total number of rolls taken and calculate the fraction of rolls that gave us each score. These fractions are our measured probability of each score. This gives us a picture of how the dice behaves when it is rolling and unobserved – proving that the dice is fair or not.

It turns out that the more rolls we make the more accurate our results will become. If you could roll a dice an infinite number of times, you could build up an exact picture of how it was rolling. This picture would perfectly describe the probability for each number facing up as a result of a roll and therefore what was happening within the roll itself. In real life, however, we can only ever get a close approximation of the exact picture, as we do not have an infinite amount of time to sit around rolling dice.

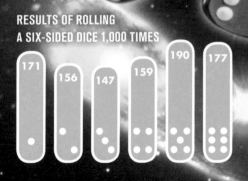

**RESULTS OF ROLLING
A SIX-SIDED DICE 1,000 TIMES**

171 156 147 159 190 177

# QUANTUM COMPLEXITY

Experiments with all types of objects in the quantum world are the same as the one I have described with the dice. We can observe the initial state a quantum object is in – like looking at the number on the dice before we roll it. The final result of some manipulation, motion, or movement through an experiment can then also be observed – like the number seen on the dice after rolling. Only from observing many quantum objects before and after an experiment can we begin to build a picture of what was going on when we couldn't observe it. Understanding all of the underlying possibilities available to quantum objects and their probabilities is just like rolling our dice many times.

However, the resulting picture we get of these properties and probabilities drawn by quantum objects is more complex than with dice. A dice represents just a single property, the score, with each number having an equal probability of becoming the result – $\frac{1}{6}$ or 16.67 per cent. Quantum objects, on the other hand, can have a number of different properties, each with different possibilities, which in turn have their own associated probabilities! For the moment let us imagine that, like the dice, we want to measure just one property of a quantum object. Now let that property be described by a picture of *Thunderbird 2*:

## ANALOGY 1

Pointillist artists build up a painting through a number of small delicate brushstrokes. Let each individual brushstroke represent an individual result from an experiment. As more experiments are run the brush strokes build up and you begin to recognise features in the picture: a booster rocket here, an aerofoil there. In a pointillist painting the detail is coarse, never continuous. The granularity of the painting, the size of the brush points, are dependent upon the number of experiments that have been conducted and the size of results sets collected.

## ANALOGY 2

Imagine that we are downloading a digital picture over the internet on an early 1990s 14.4kbits/s dial-up modem. Digital pictures are constructed from a number of pixels of different colours. Each pixel contains a small chunk of data about the picture. With just one pixel we could not tell what the picture was supposed to be. In those dark days of the Internet pictures would arrive at first as highly pixelated versions of low resolution. As more data downloaded then more of the details could be filled in. Four pixels became 16 which then became 64 until eventually we received the whole picture.

The data slowly downloading can be thought of like the results coming in from an experiment, each piece of data telling us more about the underlying possibilities and probabilities of a quantum object and bringing that picture into sharper focus. However, even now with blisteringly fast fibre optic broadband these digital pictures are still pixelated. They are a segmented, pixelated approximation of the seemingly continuous picture they represent. If we had an infinite amount of data we could build up an exacting picture describing the underlying possibilities and probabilities . . . or so it might seem.

# DUAL PERSONALITY
## WAVE PARTICLE DUALITY

Tracy Island is under attack!
Bullets are raining in through two
of the narrow windows in Tracy Villa.

On the wall behind the windows are bullet holes, which are growing in number by the minute. The bullets, of course, are most likely to fly through the centre of the windows rather than at the edges and it is here that we find the greatest intensity of gunfire. This intensity is recorded forever on the wall, as each hole registers the arrival of a bullet. We could represent the intensity of gunfire as a distribution showing the probability of being hit by a bullet. This would be highest at the centre of the windows and would then decrease out towards the edges. There would also be some bullets that have ricocheted off the window frames that spread beyond the traditional outline of the windows. Such a probability distribution would look something like this:

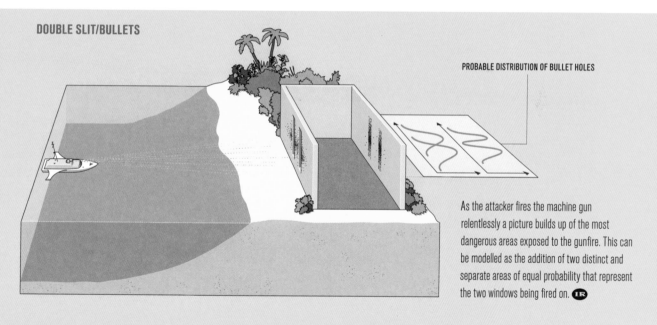

**DOUBLE SLIT/BULLETS**

PROBABLE DISTRIBUTION OF BULLET HOLES

As the attacker fires the machine gun relentlessly a picture builds up of the most dangerous areas exposed to the gunfire. This can be modelled as the addition of two distinct and separate areas of equal probability that represent the two windows being fired on. **IR**

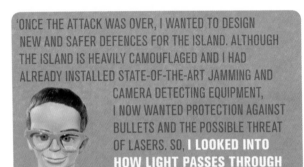

'ONCE THE ATTACK WAS OVER, I WANTED TO DESIGN NEW AND SAFER DEFENCES FOR THE ISLAND. ALTHOUGH THE ISLAND IS HEAVILY CAMOUFLAGED AND I HAD ALREADY INSTALLED STATE-OF-THE-ART JAMMING AND CAMERA DETECTING EQUIPMENT, I NOW WANTED PROTECTION AGAINST BULLETS AND THE POSSIBLE THREAT OF LASERS. SO, **I LOOKED INTO HOW LIGHT PASSES THROUGH NARROW WINDOWS. IT WAS THEN I DISCOVERED SOMETHING RATHER STRANGE.'**

# SHINING LIGHT
## SINGLE SLIT/LIGHT

As the famous Sir Isaac Newton understood it, light is made up from little packets of energy that he called corpuscles, today we call them photons. Just like individual bullets these photons cannot be divided into smaller things. When light shines through a single long narrow window photons behave, as Newton expected, in the same way as bullets. However, if we add a second window, strange things seem to happen.

## DOUBLE SLIT/LIGHT

When light shines on to a wall through two narrow windows, the pattern of light intensity is dramatically different. Where we would expect the light to behave in the same way as the machine gun bullets, instead we see that the light appears to make a pattern of fringes on the wall, with bands of bright and dark light.

PATTERN OF LIGHT INTENSITY

Tracy Island is prepared for all manner of threat thanks to Brains' knowledge of quantum physics.

Threat of a laser weapon attack is very different from that of a machine gun.

This appears to make no sense if we think, like Sir Isaac Newton, that light photons behave like bullets. The only way to explain it is to change our idea about what light is. If we think the same way as the Danish scientist Christian Huygens, that light is a wave, then suddenly an explanation of this new complicated probability distribution presents itself.

## WAVE INTERFERENCE

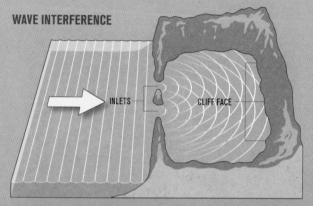

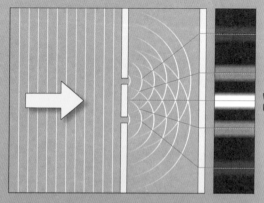

INLETS

CLIFF FACE

WAVE/LIGHT
INTENSITY

Tracy Island has two small inlets close together. The gaps in the island's cliffs act like the windows in our light experiment. We can think of the size (amplitude) of the incoming waves as a measure of the intensity. The taller the waves, the more water, and therefore the more impact on the rock of the cliff face. We can measure the size of the waves by looking at the height of buoys floating in it. In this way we can measure the intensity of the waves on the cliff face in the same way we can measure the intensity in the light experiment.

Waves spread out from each inlet towards the cliff like ripples from stones dropped in a pond. Waves from one inlet are seen to affect waves coming from the other. Where these waves overlap they interfere with each other. Instead of seeing the simple arrival of a pair of waves on the cliff face, we witness a number of spots of tall, high intensity and small, low intensity spots. High intensity is seen where the peaks of the waves coincide and create larger waves. Low intensity is seen where the top peak of one of the rippling waves meets the bottom trough of another. The pattern of intensity seen here exactly represents that seen when looking at light. OK so that wraps it up, Christian Huygens was right and Sir Isaac Newton was wrong – light is a wave. Or is it?

# PARTICLE GUN
## DOUBLE SLIT/ELECTRONS

If we were able to dial down our light source so that it was incredibly dim we would still see that light arrives in small packets of energy. Like a handgun firing just one bullet at a time into a wall, light would arrive one packet of energy at a time. So when light is emitted it seems to behave like a bullet. When travelling, however, it seems to choose the path it follows as if it were a wave. So what is going on here? This is an argument that raged between Isaac Newton and Christian Huygens and many other distinguished scientists, for centuries.

In the hands of major criminals like The Hood particle beam weapons could pose a very real threat to world security. **TB**

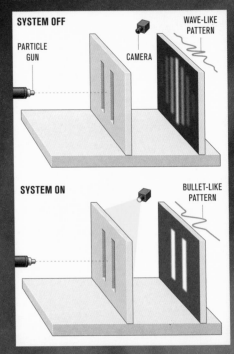

**EARLY WARNING SYSTEM**

SYSTEM OFF — PARTICLE GUN — CAMERA — WAVE-LIKE PATTERN

SYSTEM ON — BULLET-LIKE PATTERN

Newton argued that light was made from photon particles, just like bullets. Huygens, on the other hand, argued that light moved in waves, just like the water that surrounds Tracy Island. In their own ways they were both right, but ultimately wrong. It wasn't until the birth of quantum physics in the early 20th century that we understood light was something altogether different – both wave *and* particle combined.

Even if we were able to fire one photon particle at a time through the two slits, the experiment would eventually result in the 'fringe' pattern seen on our cliff face. This pattern would take a long time to build up, as just one packet of energy would arrive at a time, but eventually it would show itself. So if light behaves like particles but also like waves then what about other tiny quantum things like electrons (the very small particles that are part of an atom)? We don't see them acting exactly like bullets and we don't see them acting

exactly like waves; but they do behave like photons. For International Rescue it was good news because any defences designed to protect against laser light would also protect against electron or other particle guns.

## NOT THE END

But that was not the end of it. Brains' plan was to experiment with and develop an early warning system that could predict which slit the photons of laser light would go through. Such a system could help predict which window was in most danger of attack at any moment. The results of his experiment were astonishing. When the early warning system was turned on, he saw the photons behave like bullets; when he switched it off, the light pattern was as seen with waves on the cliff face. It seems that in the very act of detecting the light particles changed their behaviour. This is what makes quantum things so strange and ultimately amazing.

# WHAT'S THE MATTER?
## PARTICLES AND FIELDS

Quantum physics is the study of the behaviour of the smallest things in the universe. But what exactly are they?

'AT THE TIME OF HIS DISCOVERY GELLMAN WAS READING JAMES JOYCE'S NOVEL *FINNEGAN'S WAKE*. HE WAS STRUCK BY A PASSAGE THAT READ, **"THREE QUARKS FOR MUSTER MARK! SURE HE HAS NOT GOT MUCH OF A BARK. AND SURE ANY HE HAS IT'S ALL BESIDE THE MARK."**

HE LOVED THE SOUND OF THE WORD (PRONOUNCED 'KWARK') AND THE FACT THAT THE NUMBER THREE FITTED THE WAY QUARKS APPEAR IN NATURE. FELLOW SCIENTIST GEORGE ZWEIG, PREFERRED THE NAME ACE, BUT QUARK IS THE NAME THAT HAS STUCK.'

At the turn of the 20th century most thought that the atom (taken from the Greek word for 'indivisible') was the smallest particle of matter. Thirty years later scientists were able to show that atoms were actually made up of electrons, protons and neutrons. In the 1960s many experiments were conducted by smashing these particles into each other. Results showed that unlike electrons, protons in collision produced much messier debris; acting like two bags of groceries smashed together. The debris of particle collisions contained a whole zoo of new particles that scientists had never seen before. Then mathematical physicist Murray Gellman noticed a pattern in the results of the experiments, noting that the new particles could all be made from a combination of smaller particles, called quarks.

Gellman identified that quarks appeared in threes inside protons and neutrons, but only comprised two types: up quarks and down quarks, with a proton having two up and one down and a neutron having two down and one up. Further research revealed that Nature has chosen to replicate these quarks and the electron in two sets of heavier forms. Alongside these there are also three particles called neutrinos, associated with the electron and its heavier cousins the leptons. These twelve particles and the interactions between them are the fundamental constituents of matter; the building blocks that make up the visible Universe around us.

## THE STANDARD MODEL

By the mid-1970s the model of these different building blocks and our understanding of the forces that bind them together was developed into a theory called the Standard Model. Before we discuss these 'forces' we need to understand how a particle is understood in this theory.

As we have seen, when interacting a particle is a point in space and time with certain properties. In between interacting these quantum objects show wave-like behaviour. We have been thinking of these as one-dimensional waves, like waves on water (see pages 45–6), but now we need to extend the notion of a wave into all three space dimensions and also the fourth dimension of time. Such a four-dimensional wave, which also obeys Einstein's special theory of relativity, is known as a field. Fundamental particles are described through the use of

fields. In the theory every particle has an associated field and every field a particle. Encapsulated in each field are all of the properties of each particle – all of those possibilities and probabilities rolled into one.

### STANDARD MODEL OF FUNDAMENTAL PARTICLES

| | FERMIONS (THREE GENERATIONS OF MATTER) | | | BOSONS |
|---|---|---|---|---|
| MASS<br>CHARGE<br>SPIN | 2.4 MeV/c²<br>⅔<br>½<br>**u**<br>up | 1.27 GeV/c²<br>⅔<br>½<br>**c**<br>charm | 171.2 GeV/c²<br>⅔<br>½<br>**t**<br>top | 0<br>0<br>1<br>**γ**<br>photon |
| QUARKS | 4.8 MeV/c²<br>-⅓<br>½<br>**d**<br>down | 104 MeV/c²<br>-⅓<br>½<br>**s**<br>strange | 4.2 GeV/c²<br>-⅓<br>½<br>**b**<br>bottom | 0<br>0<br>1<br>**g**<br>gluon |
| LEPTONS | <2.2 eV/c²<br>-⅓<br>½<br>**ν**<br>muon<br>neutrino | <0.17 MeV/c²<br>-⅓<br>½<br>**ν**<br>muon<br>neutrino | <15.5 MeV/c²<br>-⅓<br>½<br>**ν**<br>muon<br>neutrino | 91.2 GeV/c²<br>0<br>1<br>**Z**<br>Z boson |
| | 0.511 MeV/c²<br>-1<br>½<br>**e**<br>electron | 105.7 MeV/c²<br>-1<br>½<br>**μ**<br>muon | 1.777 GeV/c²<br>-1<br>½<br>**τ**<br>tau | 80.4 GeV/c²<br>±1<br>1<br>**W±**<br>W boson |
| | I | II | III | |

# MAGNETS, ELECTRICITY AND NUCLEAR DECAY
## FORCES OF NATURE; THE ELECTROWEAK FORCE

Particles know how to interact with one another and the world around them thanks to the forces of Nature. There are four forces we know of that dictate everything. Gravity is the most familiar as this keeps our feet on the ground and the Earth orbiting the Sun. However, tiny particles have such small mass that gravity has little effect on them, so the Standard Model discards it. The three forces that do effect the quantum world are the electromagnetic, the weak nuclear and the strong nuclear forces. The electromagnetic force is responsible for electric charges and their combined magnetic fields. The weak nuclear force is responsible for nuclear power and the process that makes the Sun shine. The strong nuclear force keeps all the positively charged protons stuck together in the nucleus of every atom.

Imagine two skaters playing catch on an ice rink. Newton explained that when one skater throws the ball to the other they both start to move backwards – one through pushing and the other through catching – because they are exchanging energy through the ball. Information and energy associated with the three forces relevant to the quantum world are exchanged in a similar way between particles using an additional set of fundamental particles called bosons.

## AYE AYE BOSON

Like the other fundamental particles, these force-carrying particles are represented by a field. As in the double-slit experiment (see page 48), the wave-like fields of any two particles will change the pattern of probability of a certain thing happening.

Photons carry the information and energy of the electromagnetic force. Anything involving an electric charge or magnetic field is driven by an exchange of photons. For example, iron filings spread around a bar magnet align themselves into a pattern, which traces out its magnetic field. The field is created because most of the atoms in the magnet are aligned in one direction. This enhances the small magnetic field each charged atom has and every small bit adds up to form a large magnetic field. The atoms in the magnet announce their presence by exchanging photons that exert a force on electrons in the iron and pull them into line with the magnetic field.

In the same way the electromagnetic and weak forces are also part of a common force. It turns out that some of the weak force-carrying particles also have an

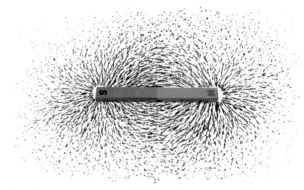

Iron filings are drawn to magnetic field lines through the persuasive exchange of photons. **IR**

electric charge. This is only possible because the weak nuclear force and the electromagnetic force come from the same origin. They are combined to form the electroweak force. The photon is one of the particles in this electroweak force. There is also a heavier version of the photon called the $Z^\circ$. Then there are the W bosons, one with negative electric charge and the other positive. These are the particles that embody the electroweak force. The combination of these two, at first glance quite different forces was made possible by a common connection. This connection was the existence of a new type of field and its associated particle – the Higgs boson.

## THE HIGGS BOSON

We have talked about heavy and light particles, but at quantum scales we do not really consider gravity as a force due to the fact that it is extremely feeble. Instead we think of the mass of particles. Mass is not weight, rather it is the measure of how easy or difficult it is to move something with a given push. Weight is the push of Earth's gravity multiplied by this mass.

Mass is given to particles through interaction with the field associated with a particle called the Higgs boson.

The more something interacts with the Higgs field, the greater the mass and the harder the particle finds moving. Conversely, the less massive a particle is then the easier it will find moving. As particles of light, photons have no mass and do not interact with the Higgs field at all, which means they travel at the upper speed limit of the Universe, what we call the speed of light.

## SNOW FIELD

Think of the Higgs field like a field of snow that extends to infinity in every direction. Imagine a skier traversing this field. Skimming over its surface, a skier would interact very little with the snow and travel fast.

If a hiker wearing snowshoes were to make their way across the field they would make much slower progress because of greater interaction between shoes and snow. The skier is like a low mass particle – a neutrino – held back very little by the snow. A snowshoe hiker is like a particle of more mass, such as an up quark, held up by greater interaction. A hiker without snowshoes would find the going extremely tough and slow and could be likened to a much larger mass particle, like a proton.

The Higgs field slows all of the fundamental particles and in doing so gives them mass. The existence of the Higgs boson particle associated with the field was first suggested by physicist Peter Higgs in 1964. Since then it has been the goal of thousands to be able to 'see' the Higgs boson and confirm this as the way in which Nature works. To do this, scientists had to kick up enough snow to be able to gather evidence that it is made from individual snowflakes. In 2012, almost 50 years after the first mention, the two ATLAS and CMS experiments took photographs of enough Higgs field snow, thrown up by the Large Hadron Collider at CERN in Switzerland, to confirm the particle's existence. In 2013 Peter Higgs and François Englert received the Nobel Prize in Physics for their part in what is currently scientists' most accurate understanding of Nature at its smallest.

# BIG BANG MACHINES
## PARTICLE ACCELERATORS

As we have already seen (see page 39) we must roll a dice many times to fully understand the underlying possibilities and probabilities of a getting a certain score. To find the Higgs interacting with fundamental particles we therefore had to look at as many particle interactions as possible. The pinnacle of a 25-year international

LHCb

ATLAS

CERN Meyrin

CERN Prévessin

SPS - 7 km

ISSE
RANCE

CMS

ALICE

LHC - 27 km

effort culminated in 2009 with the first running of the Large Hadron Collider (LHC). The 27km circumference ring accelerates protons to near the speed of light before smashing them into each other.

High voltage electronics create alternating charge differences surrounding each bunch of around 100 million protons so that there is an attractive negative charge in front of them and a repellent positive electric charge behind them. This method pushes and pulls the protons around the 27km 11,000 times per second.

They are guided to travel in a circle around the ring by magnets 125 times stronger than the magnetic field of the Earth. To keep the magnets cool enough to operate they are cooled by liquid helium, which makes the LHC magnets a whole degree Celsius colder than deep space.

All of this technology has been developed specifically for the task but has already found, and will continue to find, applications in every aspect of our lives. Particle physics has affected all of our lives from the World Wide Web to medical imaging, cancer treatment and the production of green energy.

The LHC smashes over 600 million protons into one another every second. As these two grocery bags of fundamental particles collide the debris that follows is sifted through for signs of a new as yet unseen particle. There are four highly sophisticated cameras that sit surrounding the points at which protons meet each other: ATLAS, CMS, ALICE and LHCb, which collect pictures of these collisions. They are then sent worldwide to thousands of scientists to search for elusive new particles, including the recently discovered Higgs.

This photographed 'event display' shows an example of a collision of protons in the ATLAS detector. The tracks left by particles can be seen coming from the central collision point, in the noise we can pick out the creation of a Higgs boson.

The LHC and its experiments at CERN in Switzerland.

# SEEING THE INVISIBLE
## PARTICLE DETECTORS

Particle detectors are the largest and most sophisticated cameras made by man. They do not see in the traditional sense of the word (see page 60). Instead they must reconstruct pictures of particles from the influence they have on their surroundings.

It is very much like understanding the size and speed of *Thunderbird 4* from simply looking at the waves it leaves in its wake. In the same way that *Thunderbird 4* stirs the water around it to form waves, an electrically charged particle excites electrons orbiting atoms that it passes. In their excitement these electrons either emit light, heat, or are freed entirely from the atoms they surround. The light, heat, or escapee electrons are then captured and turned into electrical signals that are then pieced together like a giant jigsaw to reconstruct a picture of what particle went where and when. From all this information we can understand the debris that results from two full grocery bags of protons colliding.

The 'A Torriodal LHC Apparatus' (ATLAS) and 'Compact Muon Solenoid' (CMS) particle detectors

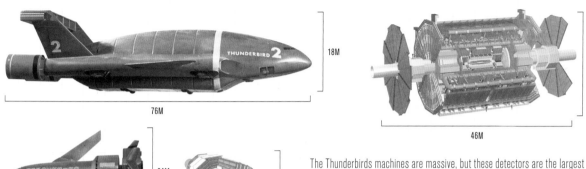

18M

25M

76M

46M

24M

15M

35M

21.5M

The Thunderbirds machines are massive, but these detectors are the largest and most sensitive cameras ever built. ATLAS (above) alone is more than five double-decker buses high and around ten double-decker buses in length. CMS (left) weighs in around the same as 75 full size blue whales, or over 1,400 African elephants. **TB**

weigh in at 7,000 and 14,000 tonnes each. ATLAS is 46m in length and 25m in height while CMS is 21.5m long and 15m tall. Each detector is made from onion like shells of different technologies. They are each capable of resolving the movement of particles smaller than the width of a human hair. Every second each detector takes tens of thousands of photographs and uses specialised electronics to choose which ones to keep and which to throw away. The reason that some must be thrown away is that there is no possible physical way of recording all the photographs. Even after reducing the number of photographs by a factor of 200,000, it still saves the equivalent of four DVDs worth of photos every minute.

These machines represent the largest and most complex technology ever built by man. They also have the most sophisticated electronics and computer systems ever designed, to handle the information they produce. Scientists then sift through the photographs to find the evidence needed to discover new science. This is the greatest international achievement in living memory.

# HOW WE SEE

The human eye is amazing. From the stars in the sky to insects on Earth, we can see things of all shapes and sizes. But what do we mean by 'seeing'? For us to be able to see we need visible light. We see stars because they radiate (send out) large amounts of visible light. Most things on Earth, living or otherwise, do not produce their own light. Instead, we see them because of the visible light from the Sun bouncing off them and into our eyes.

## HOW PHOTONS BEHAVE

Light is made up of packets of energy called photons. When reflecting, photons behave in a similar way to solid balls. Photons will bounce off a surface like ping pong balls off a table. If we wanted to 'see' something with ping pong balls we could bounce them off that object. The pattern they form when they bounce off can be used to create a picture of the underlying structure of an object. Try this yourself. If you drop ping pong balls on to an egg box the balls scatter in a number of

directions as
each one bounces off the
finer detail of the egg box's structure.
By measuring a large number of these bouncing
balls, you could build up a clear picture of what the
egg box looked like. You could see the rise and fall of the
individual egg holders, for example. This is an analogy of
how we see detail in the world around us. Photons bounce off
structures and our eyes reconstruct these into vivid pictures.
The structure of the egg box, however, can only
be resolved because the ping pong balls are small enough.
If we tried to see the fine structure of the egg box using
beach balls this method would not work. Beach balls
would not be effected by the smaller scale structure
of the egg box and would most likely bounce off it
as if it were a flat surface. The ping pong balls are
smaller than the structure they are trying
to resolve, so this works. Exactly the
same happens with photons
of light.

# WAVELENGTH

Visible light is defined not by type, as with ping pong and beach balls, but instead by a characteristic size called wavelength. The wavelength of visible light defines the colours our brains reconstruct when forming pictures from scattered photons.

Larger wavelength light is redder and shorter wavelength light is more blue. The same way different sized balls are used to measure the structure of the egg box, we can use different wavelengths of light to see different sized structures. The wavelength of visible light varies from 350–750 nanometres (billionths of a metre). The relative difference in the wavelength of visible light is small, differing by a factor of roughly two. This means that we can only improve on the smallest structure we can resolve using visible light by a factor of two. However, there is light that exists at larger and smaller wavelengths – outside the range we call visible light. Our eyes have evolved to be sensitive to these visible wavelengths only because they are the most abundant source of light produced by the Sun.

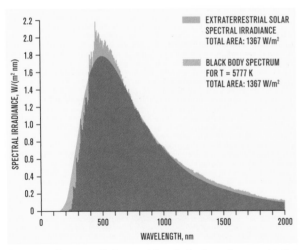

The measured intensity of light emitted from the Sun (orange) compared to that of a perfect black body of a similar temperature (grey). This shows that the Sun, though not perfect, is a good representation of a black body. **IR**

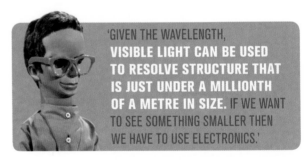

'GIVEN THE WAVELENGTH, **VISIBLE LIGHT CAN BE USED TO RESOLVE STRUCTURE THAT IS JUST UNDER A MILLIONTH OF A METRE IN SIZE.** IF WE WANT TO SEE SOMETHING SMALLER THEN WE HAVE TO USE ELECTRONICS.'

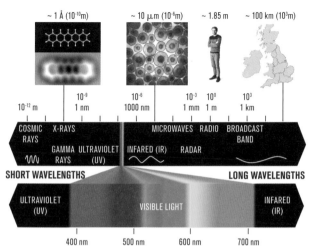

~ 1 Å (10⁻¹⁰m)     ~ 10 μm (10⁻⁶m)     ~ 1.85 m     ~ 100 km (10³m)

| 10⁻¹² m | 10⁻⁹ 1 nm | 10⁻⁶ 1000 nm | 10⁻³ 1 mm | 10⁰ 1 m | 10³ 1 km |

COSMIC RAYS   X-RAYS         MICROWAVES   RADIO   BROADCAST BAND

GAMMA   ULTRAVIOLET   INFARED (IR)   RADAR
RAYS      (UV)

**SHORT WAVELENGTHS**                    **LONG WAVELENGTHS**

ULTRAVIOLET (UV)        VISIBLE LIGHT        INFARED (IR)

400 nm     500 nm     600 nm     700 nm

Light is a name given to all forms of electromagnetic radiation, not just visible light we are used to. Radio waves, microwaves, infrared, ultraviolet, X-rays and gamma rays of decreasing wavelength are used to probe Nature at different scales. **IR**

# LIGHT, CAMERA, SCIENCE

In the classical world in which we live when we lose our keys we look for them. When we find them, their position is known to us and they do not move. In the quantum world, however, the very action of seeing something has a profound effect on whatever it is you are looking at.

To see the position of a subatomic particle – like an electron, for example – we still have to bounce photons of light off it. A photon can only resolve a position, or structure, with as much accuracy as its wavelength. We can never know the exact location of the electron because this would require us to decrease the wavelength of light to zero, or make its energy infinite in size. However, light can still locate the position of an electron to a high degree of accuracy.

What if we now wanted to measure the speed of the same electron? If we could accurately measure the position and speed of an electron we should be able to predict exactly where and when it would be moving to next.

However, the bouncing of photons off anything imparts some fraction of their energy on to whatever is being observed. Short wavelength photons have a high amount of energy and long wavelength photons less. So if we use small wavelength photons to determine an electron's position, the uncertainty of the energy we impart is large, leading to a large uncertainty in the predicted speed of the electron. The contrary is also true. With long wavelength, light will not effect the energy of the electron as much, giving us a better understanding of the electron's speed, but less about its exact location. This is called the uncertainty principle.

# UNCERTAINTY PRINCIPLE

First formulated in 1927 by Werner Heisenberg, the uncertainty principle is one of the cornerstones of quantum physics. In the quantum world things cannot be determined exactly, we can only predict a probability of possible outcomes from any measured information (see page 38). This probabilistic nature of quantum mechanics is resolved by Heisenberg's uncertainty. The fact that we cannot determine both the position and speed of a quantum object accurately through measurement means there is a limited resolution in our picture of Nature. We may understand its position with great accuracy, but only at the expense of understanding its speed.

## FACT OF NATURE

We can think of the uncertainty of some measurement as a distribution of possible speeds or position measurements. If a distribution tightens there are fewer possible measurements that can be made, and so the accuracy increases. But as we tighten one, we naturally expand the other, thereby decreasing the accuracy. Imagine two bike pumps connected by a tube; the lower the handle the more accurate we understand that property. Push one handle down and the other comes up. We could push down on both handles at the same time; we could even cool the pistons down so that the air in the system takes up less volume. But, because of that air, we can never push both handles fully down – it is a fact of Nature.

**1**

Two bike pumps connected by a tube. The lower the handle the more accurately we understand the property.

**2**

As we push the other handle down the air rushes through the connecting pipe to the second pump and pushes that handle up.

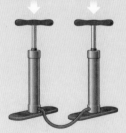

**3**

We could force both handles down, but we can never get all the air out of the system.

# PLANCK'S CONSTANT

Let's look at it another way. The resolution of a digital photograph is defined by the size of its pixels. We can seemingly keep increasing the number of pixels to gain finer and finer detail. But, no matter how hard we have tried, how perfect we attempt to make our experimental measurements we always come up against a maximum resolution. Planck's constant, published in 1900 by Max Planck, defines a very real horizon for scientific understanding; it defines the smallest possible pixel in Nature.

If we had a digital picture of *Thunderbird 3* in which we could not see (resolve) individual rivets, we could never know if a red pixel represents one, two or more rivets. The information is simply not there; it is beyond our horizon of understanding. In Nature we cannot tell what is happening below what is known as Planck's scale.

Many modern theories, such as quantum gravity and string theories, attempt to make quantum physics and Einstein's general theory of relativity work together. As yet they are still theories and focus on what could be happening in Nature beyond this minimum resolution Planck scale. These theories offer exciting explanations of how our Universe works on the most fundamental level, yet none provide any way of testing their predictions by experiment.

# QUANTUM TUNNELLING

In the classical world you throw a tennis ball against a wall and it bounces right back at you. The wall forms a barrier through which the ball cannot pass. We would never expect to find the tennis ball on the other side of the wall after throwing it. The tennis ball does not have enough energy to pass through the wall and is simply reflected, like light from a mirror. The electrons in the atoms of the tennis ball are repelling the electrons in the wall, preventing one from passing through the other.

Let us swap our tennis ball for a quantum object such as an electron. Now let us put up a barrier to the electron such as a group of atoms each surrounded by electrons. If we fire electrons at the barrier we would expect them all to be repelled and reflected back the way they came

### CLASSICAL V. QUANTUM TUNNELLING

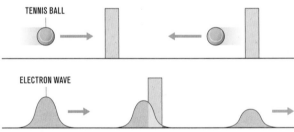

Classically, a tennis ball is not able to pass through a barrier from one side to the other (top). When quantum objects (like electrons) meet a barrier their wave-like behaviour means that there can be some small but still finite probability that they can pass right through. **IR**

by the electrons in the barrier atoms. However, what happens is that there are some electrons that make it through and can be detected on the other side.

To figure out why this happens we need to think again about the nature of quantum objects. When the ball and the wall are in contact there is a clear point at which the particle-like tennis ball ends and the wall begins. But, as we have discussed, a quantum object is particle-like only when interacting with the world around it, between interactions a quantum object travels as something like a wave. Although particles do not have the ability to penetrate a barrier, waves do. Waves are continuous and do not stop abruptly when they come into contact with a barrier. Instead waves gradually fade away through decay as they penetrate deeper into the barrier.

## NO BARRIERS

The ocean floor is a dark black place, which is why *Thunderbird 4* is fitted with an array of lights. No light reaches the strange creatures that live there. When *Thunderbird 4* starts to surface the brighter the ocean becomes, as more light is able to penetrate the shallower depths of water. The water attenuates the sunlight beating down on the ocean above. Close to the surface there is little loss in the intensity of sunlight, but this quickly decreases the deeper you dive. The same thing occurs with quantum objects in their wave-like state. The probability that they can be found at a certain location doesn't drop to zero in the face of a barrier, like a classical wall, instead the probability is attenuated by that barrier. Although it drops rapidly, dependent on the size of the barrier, there might be some small probability that the objects could be found on the other side. When a large number of quantum objects are fired at a quantum barrier some are seen to have passed right through.

# THE HUMBLE LIGHT SWITCH

A lot of modern electronics use the principle of quantum tunnelling: from the light switches used on Tracy Island and in your own home, switches on the consoles of the various Thunderbird craft to the latest computer storage. Electronics are involved in the movement of electricity. The flow of electricity is the movement of electrical energy in the form of electrons from one place to another. Metals are good conductors of electricity. At a microscopic level metals are a regular arrangement of atoms with a sea of electrons swimming freely between them. These free electrons can easily be pushed around metals when connected to an electricity supply. Electrons are pushed through metal much like water through a water pipe. An electricity supply floods a metal wire with electrons, which then have no choice but to flow along the wire.

## TUNNELLING ELECTRONS

The shiny surface of metals, wires and electrical contacts are often not pure metal. Like a cap on the end of a water pipe, the ends of wires and edges of metal contacts are naturally capped. Here the metal has reacted with oxygen in the air forming a layer of metal oxide. Aluminium is a highly reactive metal that can only be used to hold our favourite soft drinks because there is a layer of aluminium oxide coating it wherever it is exposed to the air. These metal oxides are not good at letting electrons flow as they are all used in connecting metal atoms to oxygen atoms in chemical bonds. This forms a very real barrier for electrons wishing to pass. The only reason that they can pass is thanks to quantum tunnelling. The probability of their being discovered beyond the capping of metal oxide means that a large amount of electrons are able to pass through – something classically impossible for water in a pipe. When switches are thrown and buttons are pressed these tunnelling electrons complete simple circuits.

'ALTHOUGH POWERED BY A COMPLEX ATOMIC FUSION REACTOR AND CAPABLE OF SPEEDS UP TO 5,000MPH, **I USED QUANTUM TUNNELLING TO MAKE** *THUNDERBIRD 2*'**S ELECTRONIC SYSTEMS AS SIMPLE AS POSSIBLE.**'

# FLASH MEMORY

The same process is used in computer memory. Whether it is storing the latest chapter of the book you are writing on your USB stick or temporarily holding information to be processed in RAM, we use the principle of quantum tunnelling. Such memory is made up of millions upon millions of electronics called transistors that can be thought of as small boxes. Each of these boxes can be filled or emptied of an electron. If there are no electrons in the box, then a computer reads it as a number one and if there are electrons present it reads it as a zero.

## QUANTUM TUNNELLING AT WORK

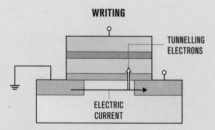

**WRITING**

TUNNELLING ELECTRONS

ELECTRIC CURRENT

A large electric voltage is applied allowing electrons enough energy to tunnel into the transistor gate.

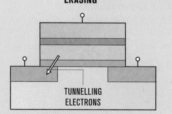

**ERASING**

TUNNELLING ELECTRONS

A negative electric voltage is applied and all electrons tunnel from the gate to an attached electrode.

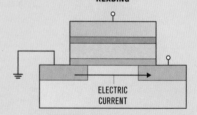

**READING**

ELECTRIC CURRENT

A small voltage is applied across the transistor; the size of current making it to the other side depends upon the number electrons stored in the gate.

Every box is sealed with a lid of the poor electrical conductor metal oxide to prevent the electrons flowing through. This lid forms a barrier large enough that the likelihood of seeing an electron on the other side is practically zero. The electron is trapped and this makes the box look different to the empty boxes. Much in the same way that we could tell if there were a present or not in a gift box by weighing it in our hands, we can tell if an electron is trapped in a transistor or not.

If we wanted to erase all the information we have stored by removing the electrons we would classically have to remove the lid. In the quantum world, however, we can do things differently. We can keep the lid on but reduce the size of a barrier that it poses to the trapped electrons. By applying a strong electric charge to the metal oxide lid you effectively reduce the barrier. The barrier is made small enough that there is now a greater probability that the electrons can be found on the other side and they tunnel through, erasing with them any information that might have been held.

## STMs

We have mentioned the limit of light in seeing individual quantum objects due to their tiny size. The principle of quantum tunnelling can offer us the most accurate picture of Nature, letting us see individual atoms placed on the surface of a metal. The size of a barrier presented to an electron depends upon the material that stands in its way. The metal oxides we have mentioned offer formidable barriers. The barrier size also depends on how thick the material is. Scanning tunnelling microscopes (STMs) use this thickness to determine the upward and downward movements of single atoms.

The microscope has a needle with the sharpest of points, which is just one

Top: A scanning tunnelling microscope set up in a lab.
Bottom: An STM image of individual atoms on the surface of blue nickel. **IR**

atom wide at its very tip. The material that is to be imaged is placed under a high voltage so that there are an excess of electrons inside it. The tip is brought within 0.4 and 0.7 billionths of a metre of the material's surface. As it scans, electrons from the surface of the material under investigation quantum tunnel into the fine tip of the microscope. As the contour of the material changes even slightly there is a noticeable difference in the flow of electrons received by the microscope. Knowing the position of the microscope tip and the electrical signal a computer can then reconstruct a three-dimensional picture of the material. The more electrons tunnelling through to the tip of the microscope, the smaller the barrier. A smaller barrier is seen when the tip is closer to the atoms being imaged.

A good STM can provide a resolution of this distance to better than 0.01 billionths of a metre, while the size of an atom is around 1 billionth of a metre. Many materials can be imaged like this and we can learn about their quantum structure. Modern STMs are also used to pick up individual atoms and move them around. This imaging and manipulation of Nature on the quantum scale shows how we can use quantum understanding to see way beyond our previous horizons.

## HOW AN STM WORKS

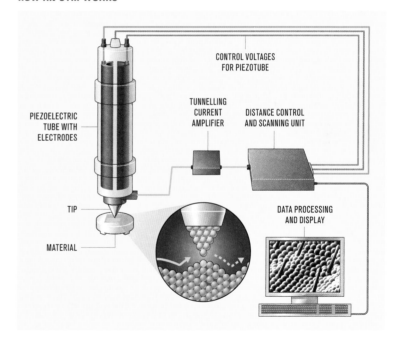

CONTROL VOLTAGES FOR PIEZOTUBE

PIEZOELECTRIC TUBE WITH ELECTRODES

TUNNELLING CURRENT AMPLIFIER

DISTANCE CONTROL AND SCANNING UNIT

TIP

MATERIAL

DATA PROCESSING AND DISPLAY

# BREAKING WAVES
## THE COPENHAGEN INTERPRETATION

Quantum physics is founded upon, and derived to fit to, experimental results. Experiment has shown us the behaviour of quantum objects, but has not allowed us a window on every aspect of the quantum world. We can still only interpret what occurs at these small scales from classical measurements of quantum objects. The physical reality of what is happening when we measure a quantum object is still a hotly debated field of research.

'QUANTUM PHYSICS DID NOT ARISE FROM NEW IDEAS ABOUT HOW NATURE IS PHYSICALLY STRUCTURED, LIKE, FOR EXAMPLE, EINSTEIN'S THEORY OF GENERAL RELATIVITY. **IDEAS WERE CREATED TO EXPLAIN HOW NATURE WAS SEEN TO BEHAVE, BUT SOON STARTED TO MAKE PREDICTIONS OF THEIR OWN.'**

Quantum objects behave as both particle and wave (see pages 43–9). A quantum object is created particle-like with well-defined classical properties such as position or mass, after this it behaves wave-like, playing out many possible scenarios with differing probabilities. At some point in the future if the quantum object interacts with its environment the spectrum of outcomes is replaced by a single classically defined particle-like outcome.

In rolling, a dice has a number of different possibilities. When the dice comes to rest, however, it is forever more observed as the number on which it lands (see page 38). This idea of the total collapse of all other possibilities other than that classically measured was put forward by Werner Heisenberg at the Solvay Conference in Brussels in 1927. When presented with this suggestion many conference attendees were uneasy about the ad-hoc nature of the interpretation.

## SCHRÖDINGER'S CAT

During an exchange of letters with Einstein (see pages 84–5), Erwin Schrödinger suggested a thought experiment that highlighted how absurd this view of quantum physics was. He suggested that the mortality of a cat placed inside a sealed box depended upon the quantum state of a radioactive atom. If the radioactive atom decayed then poison would be released to kill the cat, if it did not then the cat remained alive. He noted that after some time, thanks to his equation, you would determine that the unobserved radioactive atom would be in some mixed wave-like state (superposition) of both decayed and undecayed, rendering the fate of the cat also in some mixture of both dead and alive! Although intended

to draw attention to the absurdity of the Copenhagen interpretation, Schrödinger's cat experiment has instead proved to be a model against which this and many other interpretations have since been measured.

Having heard of the mixed fate of Schrödinger's cat, the Hood devises a devious trap. Gordon Tracy is restrained inside *Thunderbird 4* while the possible radioactive decay of an atom decides his fate; will the atom decay and turn off his oxygen or will he live to fight another day? We would have to open the airlock and find out. **IR**

# COMPETING THEORIES

The immediate disappearance of all other possible properties of a quantum object when observed troubled many. It suggested that quantum physics was a black box from which we could only observe the probability of certain outcomes. This total collapse of the wave-like nature shielded us – like an internet server's firewall blocks us from certain websites – from the complexity that existed in the spectrum of possibilities and their probabilities. The Copenhagen interpretation is the most popular view of the physical reality of quantum physics, the one taught in nearly all undergraduate physics degrees and the one we discuss most in this book. However, because of the many 'paradoxes' and worries raised about it there are a number of other interpretations that have found popularity.

## QUANTUM DECOHERENCE

Decoherence is an interpretation as follows: an observer only perceives the collapse of the wave-like nature of a quantum object; it does not completely collapse. In the Copenhagen principle the role of the observer is all but ignored. Decoherence tells us that we cannot ignore the role of the observer in the outcome that is observed. Any observer, or experimental equipment, will itself have some wave function, as it too is made from quantum objects. The interpretation then states that the apparent collapse of the wave-like nature of a quantum object being observed comes from an interaction of the quantum object's wave of possibilities and that of the observer. Any two waves can interact to form a new pattern of waves (see page 47), as the wave-like

The idea of the observed object and the observer being bound together as a single system means that not only do we have to consider the effect of the observed object on the observer, but also the effect of the observer (and his/her environment) on the observed object.

nature of a quantum object is affected by the wave-like nature of its surroundings.

In special circumstances this interaction could render the possibilities of all properties of the quantum object to zero, apart from just one. As an observer of such a thing we would perceive such a quantum object to be not wave-like but instead a particle with just a single property. After the 'measurement' the quantum object becomes part of a global combination, called superposition, with the environment's wave-like nature that 'measured' it in the first place. This is similar to a powerful jet observed to push water up into a fountain before falling back down to become indistinguishable from the rest of the water. In Schrödinger's cat experiment this would render the cat either dead or alive and not some coherent mixture of both possible outcomes.

The idea of decoherence took away the undesired idea of some unknown happenings in a black box and brought it back to understanding waves. If we knew the wave-like nature of the observer we could understand the underlying wave-like nature of any quantum object observed. There's just one problem, observers are large and classically understood objects made of trillions upon trillions of atoms each contributing their

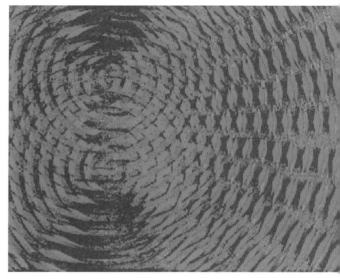

Interacting waves create a new unique pattern. Decoherence interpretation tells us these superimposed waves can be observed as measurement of a classical property of a quantum object, for example, momentum. **IR**

own influence on the wave-like whole. Even today, with modern computing power, we find it difficult to model and understand the nucleus of a single atom. Decoherence offered a way forward as an explanation, but experimentally it posed difficulties.

## MANY-WORLDS

In 1957 Hugh Everett published what he called the Universal State formulation. His idea implied that if some quantum object were to interact with this Universal State then the two waves would merge as described in the decoherence theory. The superposition of the Universal State and the quantum object's wave-like behaviour would result in all possible outcomes of the quantum object's observation to be true, that each of the outcomes was played out, but in different alternative universes. As observers we can only observe one outcome because we live in just one universe. In terms of Schrödinger's

In the Many-Worlds interpretation death *and* life are the result of Schrödinger's cat experiment and the Hood's dastardly plans. However, we only experience one outcome as the other is played out in another, undetectable universe. **IR**

cat, even before the box is opened it's fate has been sealed. It is then only a matter of peeking inside the box to see which universe we have ended up in: cat dead or alive. This idea became known as the Many-Worlds interpretation after it was popularised by Bryce DeWitt.

## WHAT DO WE KNOW?

At a fundamental level we still have a long way to go in understanding Nature. Our Standard Model of subatomic particles and the forces that guide them is proving to be amazingly accurate, however, we know it's not the whole story. Most observations of the distant cosmos, where we see the same subatomic particles and forces, fit in with the Standard Model. Yet we have not found a way of introducing gravity into it, nor can we account for some 30 per cent of the energy that we can detect in the Universe, much of which is in the form of something called dark matter. For an unknown reason there are regions of the Universe where there is a greater amount of gravitational force than there should be for the amount of stars, planets, and clouds of gas that live there. There seem to be some further unseen things that have a mass, feel and exert the pull of gravity, but do not follow the rules of the Standard Model. The theory is that these are

new, as yet unseen subatomic particles. The Large Hadron Collider (LHC) (see pages 56–7) is one of a long line of atom smashers looking for these new particles. The machine recreates conditions up to and including less than a billionth of a second after the Big Bang. To date it has seen nothing of these new particles.

If the missing energy is not locked up in new forms of particle, then it is possible that Hugh Everett's interpretation could actually be sound. Another theory suggests that the missing energy is locked up in extra dimensions within our own Universe. However, this is a catch-22 in that we have no way of detecting another universe or dimension, the very thing we would have to do to prove the theory is correct. Lack of new particles discovered by the LHC may suggest it as our only option, but right now it remains just another idea.

# QUANTUM COMPUTERS

The smartphones, computers and digital devices we use daily speak in the language of binary. They interpret a string of 1s and 0s, change the string by processing it through some program and then spit out a new string. These numbers represent single blocks of information called bits and take either the value 1 or 0. They are either on or off, 1 or 0. All our digital music, movies, pictures and communications are a special combination of bits – definite states which belong firmly in the classical world.

'IN THEORY, QUANTUM COMPUTERS THINK IN DIFFERENT WAYS, MAKING THEM FASTER THAN ORDINARY COMPUTERS, ANALYSING BILLIONS OF PIECES OF DATA AND THEREBY COMPLETING CALCULATIONS THAT HAVE PREVIOUSLY BEEN UNSOLVABLE.'

Quantum physics could be on the verge of revolutionising computing through a quantum version of the classical bit. One can manipulate quantum objects to encode the number 0 or 1, but that would simply be replacing classical bits like for like. The weird and wonderful world of quantum physics offers new opportunities when computing with quantum bits, known as qubits.

# POWER OF THE QUBIT

As we have learnt, between measurements quantum objects act in a similar way to waves (see pages 48–9). These waves represent the probability of all possible outcomes that we might end up with when we measure that quantum object in the future. Between measurements a pure qubit is therefore not defined like its classical counterpart as definitely 1 or definitely 0. Between measurements a qubit encodes different probabilities for measuring either 1 or 0 and also some addition of both. Such a combination of two or more possible outcomes is known as a superposition. Any quantum object is effectively a superposition of every possible property that it can possess. When eventually measured, however, the wave-like nature of a qubit must break down to show a classical measurement of either 1 or 0 (see page 77).

The true power of qubits does not come from standing alone but instead from their interaction with one another. The 1 or 0 state of each classical bit is entirely separate from every other bit. Aside from knowledge held by a user or a computer program, each bit itself cannot be inherently aware of the state of any other bit. A classical bit knows only if it is itself either a 1 or 0 and nothing more. A group of qubits on the other hand can be initiated in such a way that they do know something about each other. When two or more waves merge they are forever aware of each other's existence.

## ENTANGLEMENT

Ripples on a pond from two stones dropped in different locations will eventually meet. When crossing they might continue on in the outward direction they were travelling, but they would have changed slightly having interacted with another wave. After crossing the two waves will have been changed forever in some way, a memento of the other wave it passed through. In a similar way, quantum objects that have shared a common origin and are therefore aware of each other's initial wave-like nature are also inextricably linked – this is known as entanglement.

We have met entanglement before as the main thrust of the argument for decoherence as a method of wave collapse: the wave-like nature of the quantum object becomes entangled with that of an observer or measuring instrument, resulting in a perceived measured property (see pages 76–7). If a group of qubits are initialised in some common bulk way that affects them all, then these qubits are inherently entangled. After initialisation the wave-like nature of the qubits takes over with Schrödinger or Dirac's equation dictating in some calculable way how the probabilities of the qubits 1 or 0 possibility evolves. The fact that each qubit evolves in some understood way means that these entangled qubits will always have an idea of what state the other qubit(s) are in. Entanglement dictates that throughout a computer process the wave-like nature of entangled qubits are aware of each other.

This provides a new toolset from which scientists have already constructed new ways of programming possible quantum computers, offering the promise of tantalising improvements in the speed of specialist calculations.

## ALGORITHMS AND SECURITY

Posing a question to different people might require differing approaches, the same question might be asked of a computer in a number of different ways. For example: What two numbers multiply together to equal 12? One approach could be to list all numbers less than 12 and multiply each by the other. Another route could be to divide 12 by all numbers less than itself. Both are equally valid methods, both would take the same amount of time and both would result in the same answers: 1 x 12, 2 x 6 and 3 x 4. Different approaches to answering a question are known in computing as algorithms.

Qubits can behave like classical bits and so we can apply all of our current computing algorithm knowledge to quantum computers. Qubits can also act very differently to classical bits in allowing us to develop new algorithms, answering questions in different ways. Classical computers find certain questions tough to answer, requiring huge amounts of resources and time.

It is in the different way we can pose the same question that we see the true power of qubits and quantum computing. Scientists understand the limitations of quantum objects and qubits. From this understanding algorithms can be constructed that only run on quantum computers. Here is just one example:

## SHOR'S ALGORITHM

The question posed above is similar to one used in our daily lives to ensure our data is secure online. Origins of the first digital security algorithm are rooted in looking for prime numbers that multiply together to result in a given number. Prime numbers are whole numbers that can only be created in multiplication by 1 and itself. Twelve is not a prime number as we can also use the numbers 2, 3, 4, and 6 (2 x 6 or 3 x 4) to end up with it. Eleven on the other hand is a prime number because the only way to get it by multiplying two numbers is 1 x 11. Prime numbers up to 100 are: 2, 3, 5, 7, 11, 13, 17, 19, 23, 29, 31, 37, 41, 43, 47, 53, 59, 61, 67, 71, 73, 79, 83, 89 and 97. The transfer of encrypted data relies on each computer possessing a key to either encrypt data being sent or decrypt data received. This key can be generated by a pair of prime numbers. The question that needs to be answered to turn the key can be thought of as:

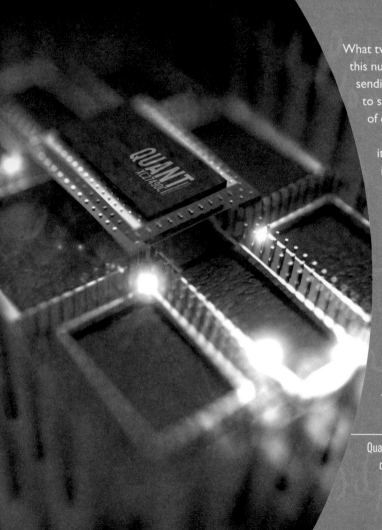

What two prime numbers multiply together to give this number? This information is known by the people sending and receiving the data; if someone wanted to steal the data they would have to find some way of calculating it.

Although this is simple for small numbers, it becomes increasingly difficult as numbers grow in size. If we were looking at numbers millions of digits in length our search would be more complicated. The most efficient method would then be to look for patterns in the numbers. This is where quantum computing really shines. The entanglement link between qubits is a global method of computation and so perfect for finding trends and patterns in lists of numbers. Mathematician Peter Shor noticed this and applied quantum thinking to the question posed above. The result was not the answer to the classical haystack/needle problem, but instead a way of using a large magnet with which to pull out the needle in question.

Quantum computers could provide a phenomenal improvement in calculation time for some of the most difficult algorithms. **IR**

# THE EPR PARADOX

After the development of the Copenhagen principle many voiced the opinion that our understanding of quantum mechanics was far from complete: three of the most prominent were Albert Einstein, Boris Podolsky and Nathan Rosen. This group published a paper, known as the EPR paper, which focused on the uncertainty principle. Heisenberg had shown that it was not possible to know both the position and momentum of a quantum object at the same time. An accurate measurement of one would result in a large uncertainty in the other. Heisenberg explained this as a result of disturbance caused by the process of measurement (see page 64).

Entanglement states that the wave-like nature of two entangled particles A and B are inextricably linked. The uncertainty principle tells us that if accurately measuring particle A's momentum then we have no hope of accurately knowing its position, but we could calculate the momentum of the as yet untouched particle B. If we then measure the position of particle B we would have an accurate idea of both its position and momentum. EPR argued that this contradicted the uncertainty principle and the Copenhagen interpretation as an understanding of quantum physics.

The paper continued to discuss how particle B might know of the measurement of particle A and visa versa. According to the Copenhagen principle the information is exchanged instantaneously at the moment of the wave collapse. This would be at a speed faster than light, an idea that contradicted special relativity, itself a pillar of modern quantum physics. A second suggestion was that there was some common link, a secret pact, based on hidden knowledge shared between the particles. This implied that the standing wave-particle formulation of quantum physics was incomplete, just an approximation of the true nature of Nature.

In 1964 John Bell suggested a number of experiments to examine the idea. These hidden variables would be a common knowledge that each particle possessed locally. If one creates many hundreds of particles in the same entangled way then these pairs of particles would each be time-sharing the same hidden variable knowledge. Measuring the properties of the pair of particles in different scenarios should show a correlated change in behaviour.

## FINDING THE LINK

The Hood has kidnapped two people and suspects them to be working for International Rescue. During the interrogation he knows he should not ask the obvious question: 'Are you working for International Rescue?' Instead he asks a series of questions about their past lives. Asking the same questions of each he can compare their answers to see if they are correlated with one another. This is the same principle used in determining the link between entangled quantum objects. The pattern of correlation can tell if the link is thanks to hidden variables (a local knowledge), or due to the quantum physics laid out in the Copenhagen principle (an instantaneous sharing of information).

Any change in a method of observation should result in some change in measurement. This change is linear in nature if the link is due to hidden variables and has a distinctly different pattern

for Copenhagen quantum physics. Many experiments looked at the correlation between electrons by measuring their spin. None of them saw evidence to back up the idea of hidden variables; instead all experiments backed up the idea of wave collapse.

## SPIN CORRELATION OF ENTANGLED ELECTRONS

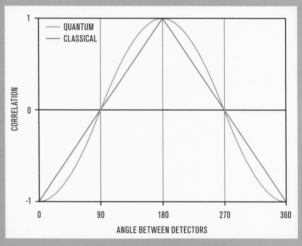

If two electrons had local hidden variable knowledge then correlation of their entanglement would follow a straight line (blue). Instead the entanglement was observed to follow that predicted by the Copenhagen interpretation (red). **IR**

Quantum physics was still left with a paradox. Einstein's theory of special relativity and subsequent experiments that verified its predictions demonstrated that there seemed to be a natural speed limit to the Universe of 300,000,000 m/s; the speed of light in a vacuum, nothing may travel faster. The wave collapse of entangled quantum objects seemed to violate this, suggesting that information is transferred instantaneously. How could that be true?

# TELEPORTATION

Quantum teleportation is not the stuff of science fiction but instead solid science fact. It is not the stuff of science fiction because it doesn't involve the transfer of stuff, quantum objects themselves, from one place to another but instead the information about the state of a quantum object. Classical information flows though optical fibres and copper cables to bring the internet to our homes. This information is sent as a stream of bits, which our computer reads and decodes before encoding another string of bits to send out once more. A bit is read from the hard drive of a computer and an electrical signal is then created representing the state 1 or 0 and sent down a wire to another computer to be read. In

quantum computing information about the state of the qubit is transferred from one location to another again without physically moving the atom or photon of light that it is encoded on. This transfer of information is what has become known as quantum teleportation.

Some people desire a secure communication between classical computers; each computer has a key code that encrypts data being sent and then decrypts data arriving. To be able to send any qubit between two computers a similar common channel must first be set up by sending two entangled pieces of information

to each quantum computer that act as keys. An entangled pair of qubits are created and one physically sent to each of two computers, which we shall call A and B. At quantum computer A the entangled qubit is used to measure the state of the qubit that is being sent. The entangled qubit has played the role of a key that has encoded the interaction between itself and the information qubit through measurement as

## HOW TO SEND A QUBIT

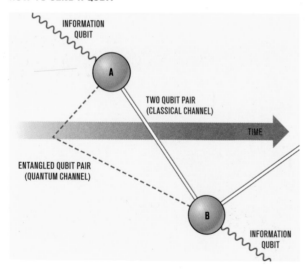

INFORMATION QUBIT

A

TWO QUBIT PAIR
(CLASSICAL CHANNEL)

TIME

ENTANGLED QUBIT PAIR
(QUANTUM CHANNEL)

B

INFORMATION
QUBIT

two classical bits. The result of the measurement is one of four classical results: 00, 01, 10, or 11. Before proceeding any further the two qubits in computer A are discarded as they are no longer useful. There is only one piece of information that can return the information qubit to its original state and that is now winging its way to computer B.

The two classical bits are sent to computer B through classical communications; thanks to the good old internet. Once received, the classical bit informs quantum computer B to manipulate the entangled qubit originally sent in one of four ways defined by the classical bits. Because these bits represent a measurement between the entangled bit and the information bit in computer A the manipulation results in an exact copy of the sent information qubit. Although we do not have the same electron, atom or photon that originally encoded the qubit we do, for intent and purpose, have the same qubit.

Although entanglement seemed to promise the possibility of faster than light exchange of information, because of the requirement of a classical link there is no violation of special relativity. Transfer of quantum information requires the transfer of classical bits that are bound by the speed of light.

# QUANTUM TECHNOLOGY

Classical bits are usually areas that have an electric charge or lack of – 1 or 0. Qubits are physically constructed from quantum objects from which we can measure certain specific quantum properties. The most popular quantum objects to use are atoms, electrons, or photons (particles of light). The quantum properties used to encode the information are either the quantum spin (see page 34) or, in the case of photons, the polarisation of light.

## POLARISED LIGHT

Light is the vibration of the electric and magnetic field, usually created from the movement or change in energy of an electrically charged particle. Most of the natural light around us has scattered off many things before reaching our eyes, which vibrates the electric and magnetic fields in many different directions (but always at right angles to the direction it is travelling). However, light can be created or made to vibrate in just one particular orientation. Such light is said to be polarised, as the vibrations are in one polar direction.

Polarised light can be created by electrons in atoms that are each aligned to produce uniform light, such as in a laser. To make polarised light from natural light you can absorb all the other components of light by passing it through an extremely fine hole, allowing just one orientation of vibration to make it through.

Modern 3D movies look 3D thanks to polarised light. Two images are projected on to the screen at the same time. Each image is made from different polarisations of light: one vibrating the electric field up-down and the other left-right at right-angles to the first. Our eyes can see all polarisations of light and so if we

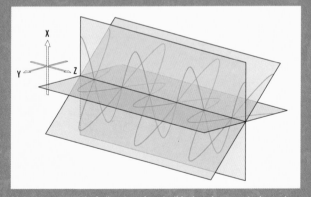

Light is comprised of vibrating electric (blue) and magnetic (green) waves. If vibrating in just one direction then the resulting light is polarised (red). **IR**

look at the screen without 3D glasses we see the two images as a blur on top of each other. Put the glasses on and suddenly we see something completely different. The left and right lenses of the glasses are designed so that they filter out and select different polarisations of light. One projected image can pass through the lens on the left and the other image through the lens on the right. Our brains then reconstruct these two images and trick us into thinking we are seeing three-dimensional pictures on the two-dimensional screen.

Some quantum computers have been constructed using similar techniques. Qubits are encoded in the polarisation of light: up-down code for 1 and left-right code for 0 (or similar). The polarisation can be controlled by specialised lasers and measured using similar methods to the lenses of 3D glasses. The light is trapped in special cavities, similar to bouncing light between mirrors.

## QUANTUM DOTS

As we discussed, particles such as electrons, and even whole atoms, have a property called spin. Spin of a particle can have different orientations with respect to some reference point, say the up-down direction (z-axis). If you are able to move the orientation of a particle's spin so that it is spinning clockwise or anticlockwise about this up-down direction then you have a way of encoding a 0 or a 1 and using it as a qubit. This is done with electrons and atoms in a number of different ways. Firstly the electrons or atoms are trapped in among other atoms in a material or in a lattice of light – the influence of surrounding atoms or laser light act as walls that contain the particle of interest. One leading technology is quantum dots: trapped electrons within a crystal of semiconductor material, for example, silicon. The position distribution of these electrons in such a material can be calculated using J. J. Thomson's plum pudding model of the atom (see pages 18–9). Although the model was not the true nature of the atom, it did help physicists when developing this quantum computing technology.

Once trapped there is more than one way of manipulating a particle qubit. The small magnetic moment within each particle acts like a tiny bar magnet. Lasers or powerful electromagnets can change the orientation of a particle's miniscule magnetic field thanks to interactions of the electromagnetic force. This is the principle behind Nuclear Magnetic Resonance (NMR) imaging in which powerful magnets are used to construct a detailed picture of the inside of our body. The trapping and manipulation of atoms and electrons allows them to be encoded as qubits.

Practically realising groups of qubits to form a quantum computer is a major driving force in modern physics and engineering. With trillions of electrons and atoms in tiny regions of materials the biggest challenge is not finding the material from which to make qubits, but instead protecting them. Qubits need protection because

if they come into contact with the outside world then the information they are encoding is lost from the computer, practically or definitely depending on which philosophy of quantum measurements you prefer (see pages 74–9).

## MAGNETIC RESONANCE IMAGING MACHINE

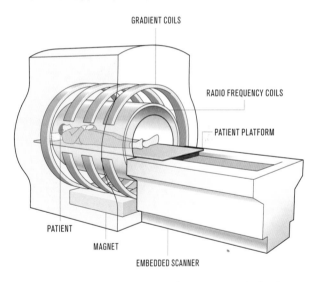

GRADIENT COILS

RADIO FREQUENCY COILS

PATIENT PLATFORM

PATIENT

MAGNET

EMBEDDED SCANNER

MRI can be used to image tissue by tweaking the tiny magnetic field of each individual atom in your body. The powerful magnets in the scanners can also be used to manipulate and read out qubits encoded in the spin of atoms or electrons. (IR)

# SUPERCONDUCTIVITY

Quantum objects may be small but they can have profound effects on the world at the human scale. MRI scanners, maglev trains and the Large Hadron Collider all require quantum behaviour to operate. At the microscopic level an electric current moving through a wire involves the flow of electrons.

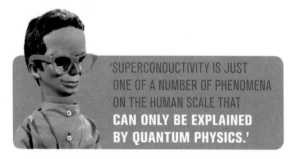

'SUPERCONDUCTIVITY IS JUST ONE OF A NUMBER OF PHENOMENA ON THE HUMAN SCALE THAT **CAN ONLY BE EXPLAINED BY QUANTUM PHYSICS.**'

The atoms of a metal are fixed in a regular arrangement surrounded by a 'sea' of electrons. When a voltage is applied they are able to move in the direction the voltage dictates. If we think of electrons and atoms like different sized balls the electrons could collide with the metal atoms. These collisions hinder their movement as they scatter in new directions. This is known as electrical resistance, with some metals having a higher resistance than others.

## FREE MOVING ELECTRONS

Atoms of a metal are fixed in a regular arrangement but each vibrate and jiggle about their position. The energy of these jiggling atoms is measured as a temperature; the higher the temperature the more things jiggle. More jiggling means an increased chance of collision and therefore a higher electrical resistance. If metals are cooled then the resistance will decrease because the jiggling becomes less. If we were to cool a metal down to the coldest possible temperature, minus 273.15°C – otherwise known as absolute zero – then the jiggling would stop. At this point we would have reduced the resistance to its lowest value possible in this classical picture of electricity. There would still inevitably be some collisions between electrons and atoms as the atoms still present a real physical barrier.

In 1911, the Dutch physicist Heike Kamerlingh Onnes was investigating the resistance of mercury metal at very cold temperatures when he discovered something amazing. At a temperature of 4.2° above absolute zero, the resistance of mercury suddenly dropped to zero. It seemed that despite our classical understanding of electric currents, the electrons were somehow able to travel totally unhindered.

To explain this we again have to think about electrons not as solid particles but as quantum objects. At this critical temperature each free electron forms a strong link with another one. This pairing of electrons, known as Cooper pairs, arises not from the particle-like nature of electrons but from their behaviour as a wave. The wave-like nature of electrons interacts with the wave-like nature of the regular arrangement of atoms, combining to create an attraction between pairs of electrons, which then flow through the metal together.

# MAGLEV

Using magnetic levitation, a vehicle can travel along rails using magnets to create lift and propulsion, reducing friction and allowing higher speeds – perfect for the Pacific Atlantic Monorail in *The Perils of Penelope*.

While the Pauli exclusion principle (see page 36) tells us that no two electrons can occupy the same space, Cooper pairs can occupy the same space within the metal and effectively travel through the gaps between metal atoms without ever bumping into one. This is known as superconductivity and is essential for technologies that require large electrical currents to flow in order to create huge magnetic fields, some large enough to levitate an entire train.

# BEYOND QUANTUM

In 1894 the American physicist Albert Michelson commented: 'The more important fundamental laws and facts of physical science have all been discovered ... the possibility of their ever being supplanted in consequence of new discoveries is exceedingly remote.'

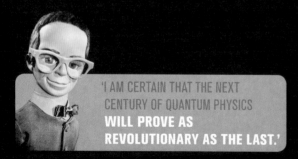

'I AM CERTAIN THAT THE NEXT CENTURY OF QUANTUM PHYSICS **WILL PROVE AS REVOLUTIONARY AS THE LAST.**'

But, six years later, Lord Kelvin noted in a lecture that there were 'two clouds on the horizon'. One was the discovery that light travels at a constant speed no matter the speed of the observer, the result of an experiment conducted by Michelson and his colleague Edward Morley. This cloud would burst into an entirely new field of science through Einstein's theory of relativity. The second was the failure to explain the observed black body radiation, and from this came the field of quantum physics.

In just over 100 years our knowledge of the quantum world has revolutionised science and technology. Understanding the structure of electrons in an atom has given us a clearer understanding of the creation and evolution of the Universe. The transistor allowed us to build computers that dominate the modern world. X-ray, NMR, CT and other medical imaging have given doctors the tools to save countless lives. The field of quantum computing promises to bring about a paradigm shift in the tools we possess for the century ahead.

## KNOWN UNKNOWNS

Would we now think Michelson's prophecy to be true? Not by a long shot. We have an incredibly long list of known unknowns. The expansion of our Universe seems to be accelerating, driven by some unknown dark energy. There is a lot of additional gravitational mass that we cannot see, determining the size and distribution of galaxies in the night sky, yet we do not know why. The heart of the most extreme environments in Nature are yet to be explained; what is in the centre of a black hole and what happened in that first tiny fraction of a second after the Big Bang? There is a very real understanding that quantum physics and general relativity alone cannot give us these answers.

Scientists have measured and explained the behaviour of this strange world to stunning degrees of accuracy. There are still pieces of the quantum puzzle to solve before the full picture is complete, but it is understood that quantum physics is limited by the small Planck scales in time and space. This firewall is shielding us from a more fundamental understanding about why quantum things behave the way they do. Beyond this lies uncharted territory, where theorists can speculate as to the very fabric of Nature. Is everything a manifestation of vibrating shapes as string theorists believe? Can dark matter be the result of as yet undiscovered subatomic particles? To answer these questions will most likely require another revolution in our interpretation of Nature, one that again spawns a whole new field of science. Through science we have interpreted the inner workings of Nature, yet it is in appreciating the limitations of this knowledge that we show our true understanding.

# INDEX

# ACKNOWLEDGEMENTS

The author would like to thank his family for their love, support, and
encouragement at every milestone of life; Emily for your love and patience
during all those evenings and weekends; and Ryan for the physics proofing.

## PICTURE CREDITS

All images supplied by ITV except the following: page 17 Bundesarchiv Bild;
pages 56, 57, 59 (both) CERN; page 72 (bottom) image originally created by
IBM Corporation; page 72 (top) Paul Logan, Courtesy of the University of Texas
at Arlington; page 58 Keith A. McNeill (www.keithmcneill.pwp.blueyonder.co.uk);
pages 20–1, 23, 37, 39 (background), 60 (left), 95 NASA; pages 9
(Pakhnyushchy), 11 (background) (Albert Barr), 16 (Tatiana Popova),
26 (Vadim Sadovski), 34 (picturepartners), 39 (dice) (stockphoto-graf),
46 (Lemonakis Antonis), 52 (Triff), 53 (imagedb.com), 54–5 (Orietta Gaspari),
61 (Phatic-Photography), 71 (Anteromite), 81 (Dmitry Naumov),
83 (Mclek), 84 (welcomia), 90 (Carlos Yudica) plus image of retro TV set used
throughout (Antonio-BanderAS) shutterstock.com; pages 48–9 Andrea Pacelli;
page 11 (inset) UK Space Agency; pages 28, 29 (Falcorian) Wikipedia.